Preface

I0505060

This text would be a useful introduction to biotechnology and genetic engineering courses. Advances in biotechnology and genetic engineering are due to the wealth of knowledge accumulated about the genetic material of human and other species. The first part of the text describes chemical structures of DNA and RNA and explains how the structure of nucleic acids relates to their functions with respect to DNA replication and protein synthesis. The codon usage bias is explained in this unit together with explanation of practical applications of codon usage bias in gene expression. This unit concludes with description of different types of single point gene mutations and their possible consequences.

Unit two begins with a description of gene regulation in bacteria. Methods used by bacteria to control gene expression at local level using operons and regulons. The control of gene expression at global level in bacteria is described, and the mechanisms a stimulon or modulon employ to regulate gene expression at global level in prokaryotes are outlined

Unit three begins with gene regulation in complex eukaryotes. The role of chromatin structure in gene regulation in complex Eukaryotes is described with respect to gene silencing and gene activation in differentiated tissues of complex Eukaryotes. The last part of this unit describes the role of epigenetic in altering chromatin structure and in gene activation or gene silencing.

Unit four Unit two describes recombinant DNA techniques used in the construction of main types of recombinant DNA expression systems and their application in the production of recombinant proteins

1

Table of Contents

1

Nucleic Acids

1.1-Introduction

Nucleic acids are molecules that cells use to store, transfer and express genetic information. In a cell, nucleic acids are represented by two forms DNA and RNA. The DNA stores and replicates genetic information and RNA is used to translate that information into proteins.

1.2-DNA

Deoxyribonucleic acid (DNA) is a nucleic acid that stores and transfers genetic or inherited instructions used in the development and functioning of all living organisms and many viruses.

Chemical structure of DNA

DNA is a duplex molecule of two polynucleotide strands. Each polynucleotide is made of chemical building blocks called nucleotide.

A DNA nucleotide is made up of 5-carbon sugar (deoxyribose), a phosphate group and one of the following four nitrogenous bases; (adenine (A), guanine (G), (cytosine (C) and thymine (T).

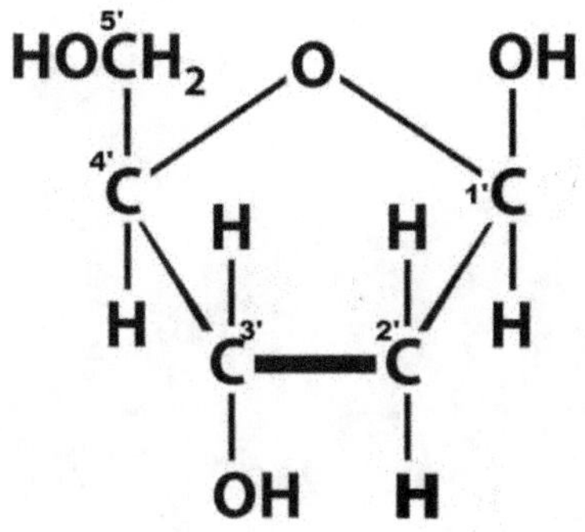

Figure 1.1. Chemical structure of deoxyribose sugar.

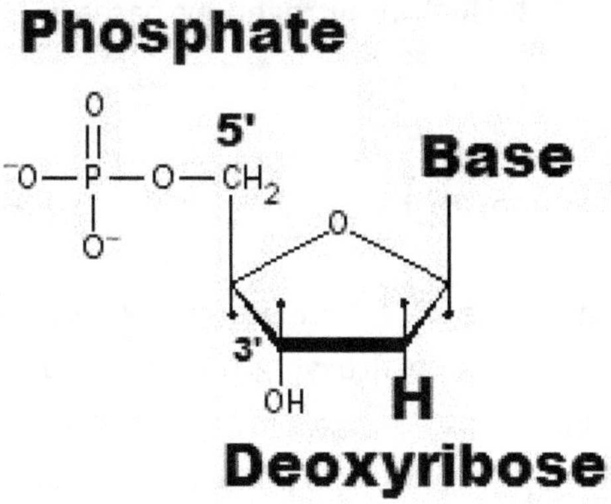

Figure 1.2. Chemical structure of a DNA nucleotide

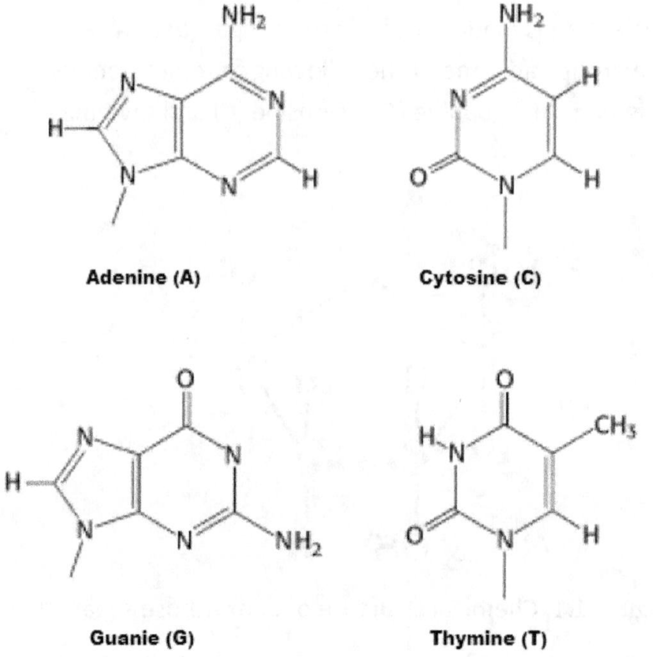

Figure 1.3. The four nitrogenous bases of DNA

DNA synthesis (Building a polynucleotide)

The first nucleotide has phosphate group on carbon number 5' on the sugar ring, and a hydroxyl group on the 3' carbon. A phosphate group at position 5' from a second nucleotide is attached to the sugar molecule in place of the OH group on the 3' carbon of the first nucleotide. A covalent bond called phosphodiester bond links two nucleotides.

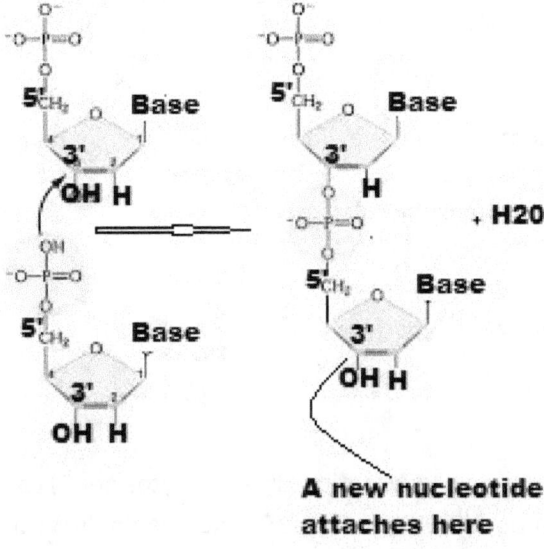

A new nucleotide
attaches here

Figure 1.4. Formation of phosphodiester bond, nucleotides synthesis from 5' (five prime) to 3' (three prime).

The first nucleotide has a free phosphate group which is known as the 5'end.   So DNA synthesis will have a phosphate group on carbon number 5' on the sugar ring in the first DNA nucleotide and a free hydroxyl group on carbon number 3' of the last nucleotide in a polynucleotide strand. Hence DNA synthesis occurs from 5' to 3'. The bond between the phosphate group and the sugar in polynucleotide molecule is called phosphodiester bond.   The above reaction is called condensation because two molecules joining together with the loss of one, in this case with the loss of water.

The base sequence or nucleotide sequence is defined as the order of nucleotides along the chain from 5' to the 3' end. The polynucleotide strand of DNA could be represented horizontally with only the sequence of nucleotides stated as in A C G C T A etc; direction is shown by indicating the 5' to 3' orientation

When two nucleotides join, they form a dinucleotide, when fewer than 20 join they are called oligonucleotides, when more than 20 join, they are called polynucleotides.

The Double-Helix Model

DNA consists of two polynucleotide strands. Cytosine (C) can form three hydrogen bonds with guanine (G) and adenine (A) can form two hydrogen bonds with thymine (T).

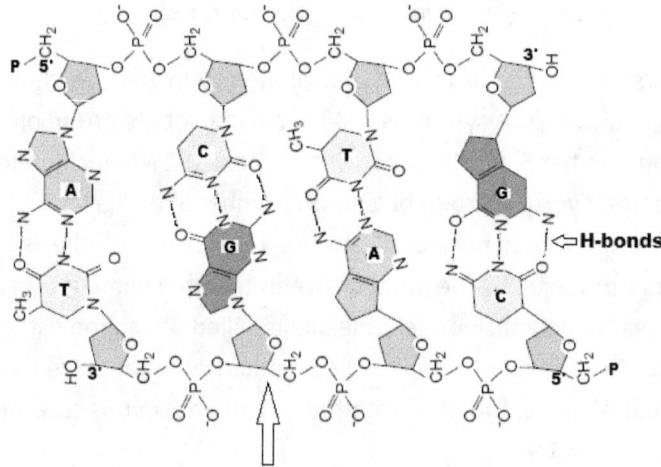

DNA backbone (alternating sugar and phosphate)

Figure 1.5. Covalent phosphodiester bonds in DNA backbone and H-bonds between complementary DNA strands.

Polynucleotides form H-bonds between their complementary bases, one strand starts with 5' to 3' and the other 3' to 5' so the

8

two strands are called complementary and ant parallel. Complementary means if you know the sequence of bases in on strand you can predict the sequence of bases on the other strand.

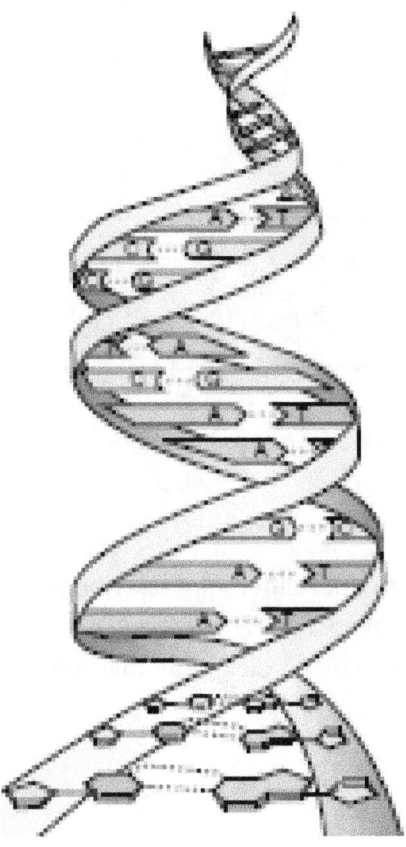

Figure 1.6. DNA structure looks like a spiral staircase with the sugar and phosphate on the sides and the steps of the staircase made from the base pairs.

Hydrogen bonds are much weaker than covalent bonds, such weak bonds are crucial to biochemical systems; they are weak enough to be reversibly broken in biochemical processes, yet they are strong enough, when many form simultaneously, to help stabilize specific structures such as the double helix.

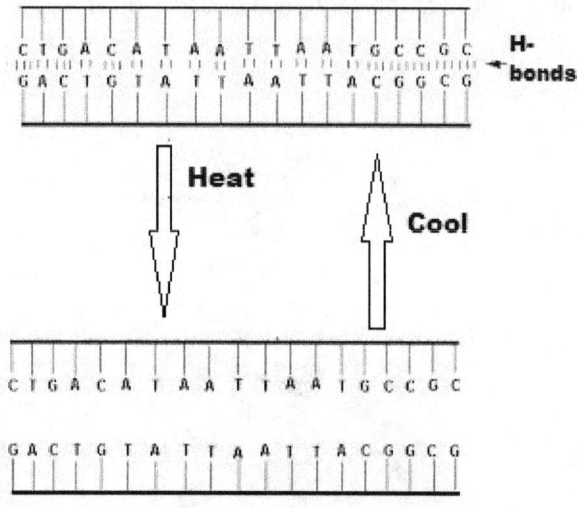

Figure 1.7. Breaking and reforming H-bonds between DNA base pairs

The sequence of the four different deoxyribonucleotides function as a four- letter alphabet, this sequence serves as a digital code that specifies the earth's biological diversity, so DNA stores genetic information as a linear sequence of bases.

1.3-RNA

Ribonucleic acid (RNA) is the second nucleic acid. It is a single stranded molecule of polynucleotides, each nucleotide is composed of ribose sugar, a phosphate group and one of four different nitrogenous bases guanine (G) cytosine (C), adenine (A) and uracil (U).

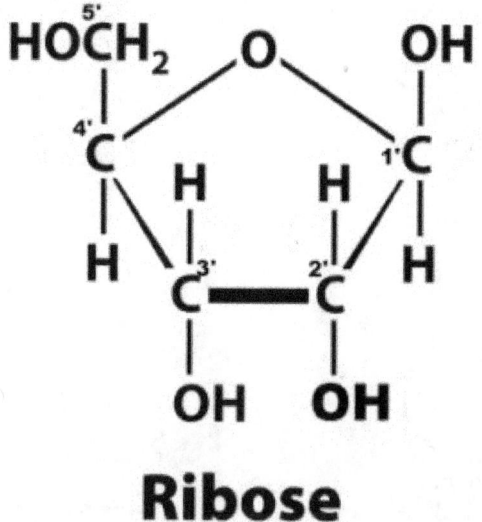

Ribose

Figure 1.8. Chemical structure of ribose sugar

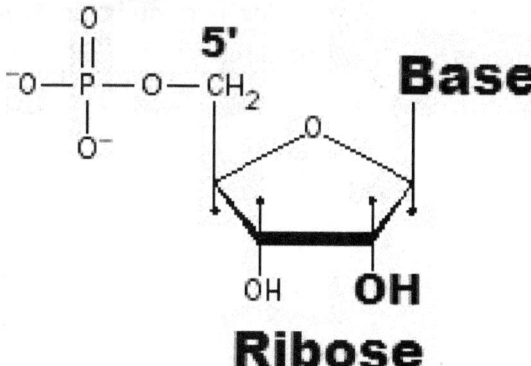

Phosphate

Base

Ribose

Figure 1.9. Chemical structure of RNA nucleotide.

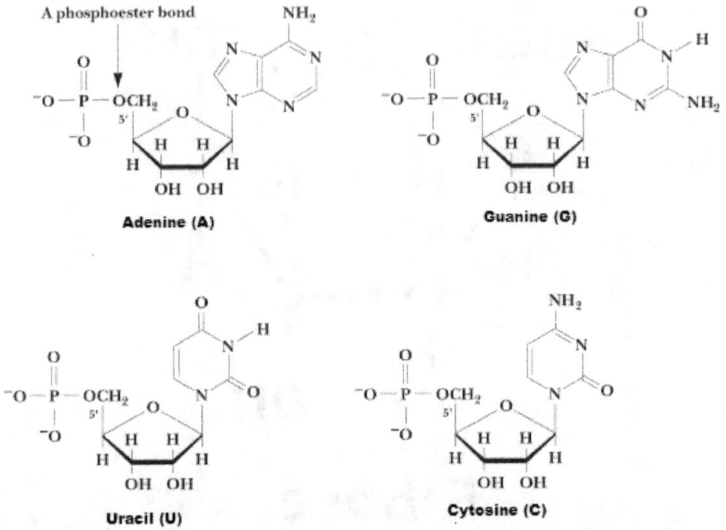

Figure 1. 10. RNA nucleotides

The links that hold the nucleotide together are the same as in DNA, a phosphate links the 5' position of ribose to the 3' position on the ribose of the neighbouring nucleotide, so RNA strand is always written as 5' to 3' orientation.

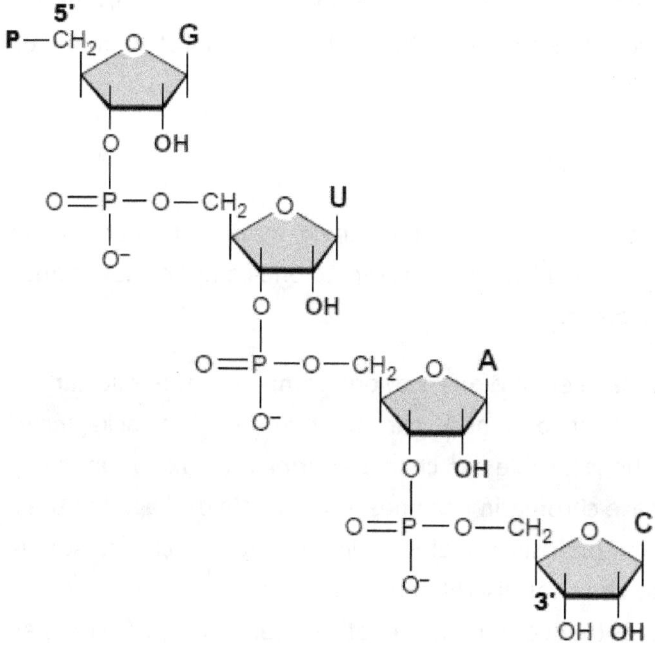

Figure 1.11. Covalent bonds between sugar and phosphate, forms the RNA backbone

The main differences between RNA and DNA are; the sugar in RNA is ribose, rather than deoxyribose, RNA contains uracil in place of thymine. Chemical structure of thymine is very similar to uracil and RNA is generally single-stranded and not double-stranded like DNA.

RNA molecules are mostly single- stranded but can also have base-pair structure. So in RNA the important base pairs are adenine (A) pairs with uracil (U) and guanine (G) pairs with cytosine (C).

Frequently such intermolecular folding is critical to RNA function example transfer RNA. If an RNA strand and a single DNA strand have complementary base sequences they can form RNA–DNA

13

hybrid double helix, when conditions are appropriate, U in RNA pairs with A in DNA, such hybrid pairing is important in DNA replication. It is also essential tool in the laboratory manipulation of nucleic acids.

1.4-Genes and genomes

The genome comprises all the genetic material that an organism possesses. In Eukaryotes it refers to one complete set of nuclear chromosomes.

The human genome refers to one complete set of nuclear chromatins, it contains 46 chromatin fibres. The packaging of chromatin is flexible and changes during the eukaryotic cell cycle, interphase chromatin becomes more tightly packaged at the time of division (mitosis or meiosis), when individual chromosomes become visible as discrete entities. Chromosomes; are the structures that contain the genetic material, they are complexes of DNA and proteins enclosed within the nucleus. Eukaryotes possess a mitochondrial genome.

Gene is the basic physical and functional unit of heredity. It consists of a specific sequence of nucleotides at a given position on a given chromosome that code for a specific protein (or, in some cases, an RNA molecule). A gene has starting point and end point; it has to be read in the correct way.

The human genome contains between 25,000 and 30,000 genes, but there are only 23 pairs of chromosomes. This means each chromosome carries many genes. The DNA sequence of the human genome has been determined, some of the genes have been identified, but many of the genes have not been identified yet, so the number of genes contained in the human genome is a current estimate.

Bacterial genome:

Bacterial genome is typically a single circular chromosomes example the chromosome in Escherichia coli = 4.6 million base pairs. The bacterial chromosome is usually concentrated in a specific clear region of the cytoplasm called nucleoid.   E.coli chromosome encodes approximately 4300 genes.

Besides the chromosomes many bacteria may also carry extra chromosomes genetic elements in the form of small circular and closed DNA molecules, called plasmids. They generally remain floated in the cytoplasm.

Viral genomes:

The chromosomal material of viruses is DNA or RNA which adopts different structures; it is circular when packaged inside the virus particle.

1.5-Origin of DNA replication

DNA structure helps explain how it duplicates. DNA replication is the process by which the original DNA molecules are copied or replicated for delivery into new cells with each cycle of cell division. If a DNA molecule is separated into two strands, each strand can act as the template for the generation of its partner strand.
DNA replication is semi-conservative i.e. it produces two copies that each contain one of the original strands and one new strand, thus the original DNA strands are used as templates for the synthesis of new strands.

15

Each strand of the double helix has all the information needed to reconstruct the other half by the mechanism of base pairing. Because each strand can be used to make the other strand, the strands are said to be complementary.

DNA replication in most prokaryotic cells starts from a single point (origin of replication) and proceeds in two directions until the entire chromosome is copied. Bacterial chromosomes have a single origin of replication, origin of chromosomal replication (Oric).

Eukaryotic cells have much more DNA. Nearly all of it is contained in chromosomes, which are in the nucleus. In eukaryotic cells, replication may begin at dozens or even hundreds of places on the DNA molecule, proceeding in both directions until each chromosome is completely copied.

The process of DNA replication is summarized here

DNA helicase enzyme attaches to the DNA at the origin of replication, which is the place where the process of DNA replication begins and moves bi directionally by unwinding the DNA double helix by breaking the hydrogen bonds between the complementary bases. The action of DNA helicase results in Y shaped fork in the molecule which is called replication fork.

During DNA replication two new strands of DNA are replicated or copied. The first one is called the leading strand. This is the strand of DNA which runs in the 5' to 3' direction toward the fork, and it's able to be replicated continuously by DNA polymerase. The other strand is called the lagging strand. This is the strand which runs in the 3' to 5' direction toward the fork, and it's replicated discontinuously. Several steps are involved in the replication of the lagging strand which means it is not as fast as the replication of the leading strand.

Leading strand

One RNA primer is made at the origin of replication by RNA primase. RNA primer is a short strand of RNA that initiate DNA replication. DNA polymerase attaches nucleotides in a 5' to 3' direction as it slides toward the opening of the replication fork.

DNA polymerases catalyze the formation of a phosphodiester bond efficiently only if the base on the incoming nucleotide is complementary to the base on the template strand. Thus, DNA polymerase is a *template directed enzyme* that synthesizes a product with a base sequence complementary to that of the template. DNA polymerase replaces RNA primer with DNA nucleotides.

Lagging strand

Synthesis is also in the 5' to 3' direction; however it occurs away from the replication fork, many RNA primers are required.

DNA polymerase uses the RNA primers to synthesize small DNA fragments; these are termed Okazaki fragments after their discoverers. DNA polymerase removes the RNA primers and fills the resulting gap with DNA. It uses a 5' to 3' exonuclease activity to digest the RNA and 5' to 3' polymerase activity to replace it with DNA, after the gap is filled a covalent bond is still missing. DNA ligase catalyzes the formation of a phosphodiester bond thereby connecting the DNA fragments.

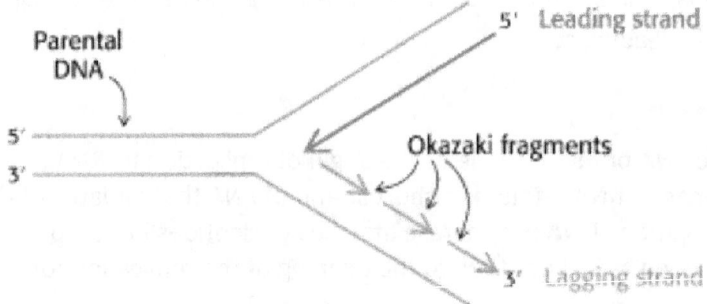

Figure 1.12. DNA replication of leading and lagging strands

1.6-Gene Transcription

Transcription (scribe) = (to write), which means to transfer data from one form to another in this case, the data are copied from DNA language to RNA language. As a result of DNA transcription, a DNA molecule is expressed in the form of the related nucleic acid (RNA).

As a result of DNA transcription several kinds of RNA are transcribed three main types involved in protein synthesis, mRNA, rRNA and tRNA in addition to many other RNAs. mRNA carries the blue print for making the polypeptide or protein; tRNA and rRNA are used later for the actual process of assembling amino acids.

Genetic information is carried on only one of the two strands of the DNA. This is known as the coding or sense strand. The other strand is called template strand.

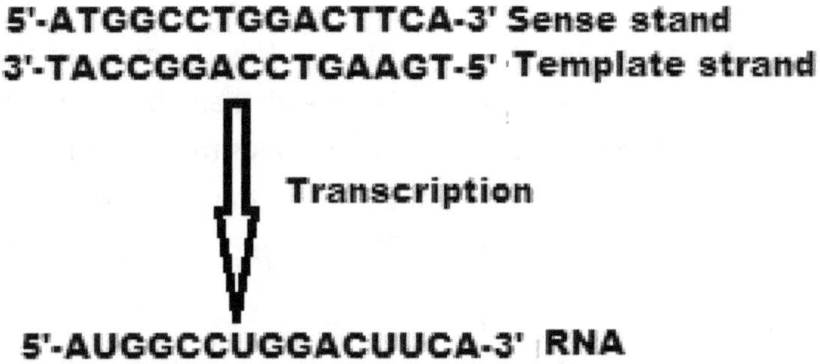

5'-ATGGCCTGGACTTCA-3' Sense stand
3'-TACCGGACCTGAAGT-5' Template strand

Transcription

5'-AUGGCCUGGACUUCA-3' RNA

Figure 1.13. Transcription of the template strand of DNA

In order to extract information from the DNA the cell uses complementary base-pairing to make a complimentary copy of the template strand as RNA strand. The orders of nucleotides A, T, G and C, on the existing DNA chain, the template determine their order on the RNA chain being transcribed. So transcribed RNA strand will be identical to the sense strand of the DNA, with (U) replaces (T).

Prokaryotic Transcription

In prokaryotes, RNA synthesis takes place in the cytoplasm.

In prokaryotes: a single RNA polymerase accounts for the production of all types of RNA.

Promoters are specific sequences in the DNA molecule that show RNA polymerase exactly where to begin making RNA. Terminator is another DNA sequences downstream of the gene, it is stop signal for transcription.

Transcription in prokaryotes can be divided into 3 stages:

1-Initiation

A promoter region upstream from gene. Upstream: mean that it is closer to the 5'end of the DNA strand than the gene. RNA polymerase recognize the promoter and binds to it, once initiation is done then RNA polymerase moves down the length of the gene and transcribes it into a strand of RNA.

2- Elongation

RNA polymerase moves along the template strand of DNA making RNA, putting in U complementary to A. Proofreading: correct incorrectly inserted base as it moves a long backtracking and removing incorrect base(s)

3-Termination

A DNA sequence downstream of the gene causes RNA to dissociate from polymerase and DNA.

Eukaryotic Transcription

In eukaryotes, RNA is produced in the cell's nucleus
Eukaryotes use three distinctive kinds of RNA polymerases to make three different types of RNAs, messenger RNA, ribosomal RNA and transfer RNA.
Transcription in Eukaryotes can be divided into three stages

1-Initiation; the signals initiating transcription in eukaryotes are generally more complex than those in prokaryotes. Transcription factors and RNA polymerase bind together in the promoter region and starts transcription
2-Elongation; Similar to prokaryotes but in the nucleus. Initial mRNA in the nucleus contains transcribed introns and exons. Introns: non-coding interrupting DNA segments. Exons: coding segments of a gene.

3-Termination; Transcription ends with untranslated region. Transcription proceeds well beyond the 3'end of the gene. Specific types of nucleotide sequences signal transcription termination and release of the completed RNA strand.

Post-transcriptional Modification

1. 5' Cap Modified guanine added to 5' start of mRNA facilitates the binding of ribosome to messenger RNA, thereby increases efficiency of translation and protects against degradation.

2. The transcript is then cleaved about 20 nucleotides past a specific sequence 5'- AAUAAA-3 that occurs after the end of the coding region. At the DNA cleavage site now the 3'end of the RNA anywhere from 50 to 200 adenine nucleotides (as) are added to form what is called a poly A tail thus the functional messenger RNA carries a non coding tail that was not present in the gene.

3. RNA may be "edited" before it is used. Portions that are cut out and discarded are called introns. The remaining pieces, known as exons, are then spliced back together to form the final mRNA. Splicing to remove introns done by spliceosomes: and small nuclear RNA (snRNA + proteins)

4. Nuclear export of mature mRNA to the cytoplasm, for translation.

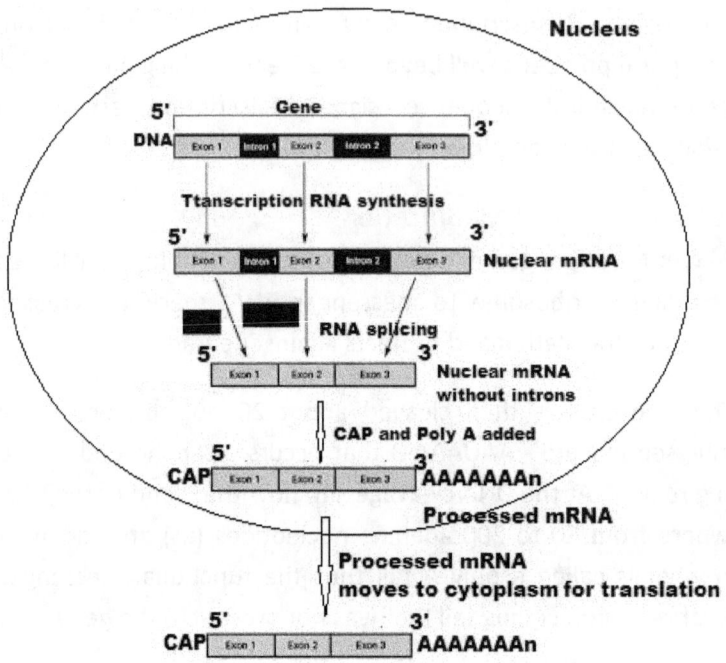

Figure 1.14. Transcription in eukaryotes

N.B: guanine nucleotide (specialized) called 5'cap and poly A tail are extra modifications to mRNA strand, they do not code for proteins.

Messenger RNA Contains Start and Stop Signals for Protein Synthesis.

1.7-The universal genetic code

The genetic code describes the relationship between the sequence of bases in mRNA and the corresponding amino acid sequence that it encodes. DNA codes for mRNA and mRNA is translated to proteins.

There are 20 different amino acids commonly found in protein in all living things. Each amino acid consists of a carboxylic acid group (COOH), an amino (NH2) and one of twenty functional groups.

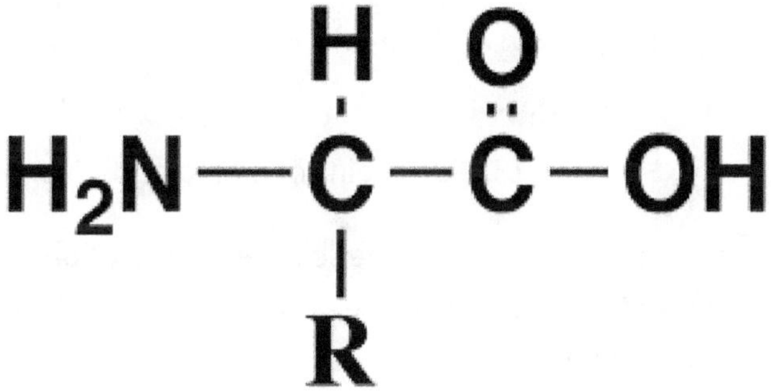

$$H_2N-\underset{\underset{R}{|}}{\overset{\overset{H}{|}}{C}}-\overset{O}{\underset{||}{C}}-OH$$

Figure 1.15. General chemical structure of amino acid

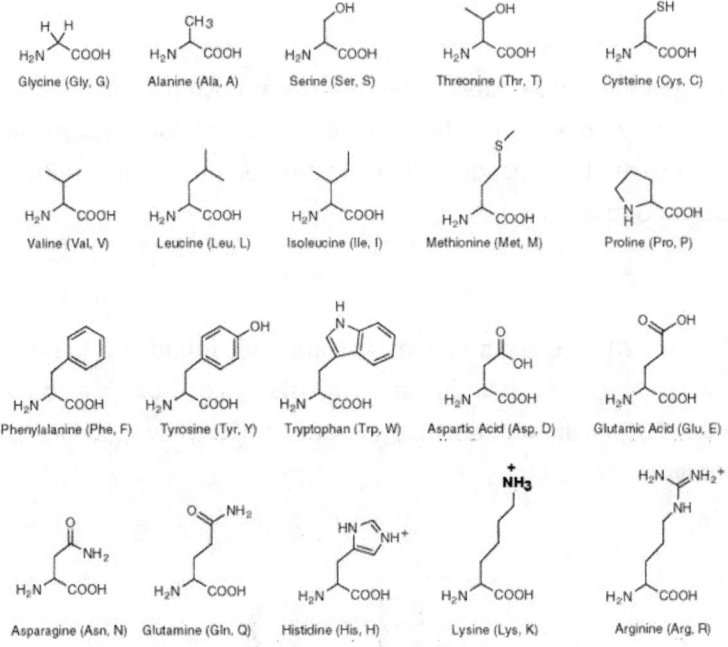

Figure 1.16. Side chains of amino acids

Codon is the basic unit for the genetic code. Each codon has three adjacent-nucleotides in the sense strand of DNA chain or its messenger RNA copy.

Second base

First base	U	C	A	G	Third base
U	Phe	Ser	Tyr	Cys	U
	Phe	Ser	Tyr	Cys	C
	Leu	Ser	STOP	STOP	A
	Leu	Ser	STOP	Trp	G
C	Leu	Pro	His	Arg	U
	Leu	Pro	His	Arg	C
	Leu	Pro	Gln	Arg	A
	Leu	Pro	Gln	Arg	G
A	Ile	Thr	Asn	Ser	U
	Ile	Thr	Asn	Ser	C
	Ile	Thr	Lys	Arg	A
	Met	Thr	Lys	Arg	G
G	Val	Ala	Asp	Gly	U
	Val	Ala	Asp	Gly	C
	Val	Ala	Glu	Gly	A
	Val	Ala	Glu	Gly	G

Figure 1.17. A codon table sets out how the triplet codons in mRNA code for specific amino acids.

The information in a gene on the coding strand is read in the direction from 5'end to the 3'end.

The genetic code contains 54 triplets of adjacent-nucleotides in mRNA or sense strand of DNA, 61 of these possible triplets each encodes only one amino acid one of these triplets, ATG in DNA or the equivalent AUG in mRNA has a dual function, it encodes the amino acid methionine and it also marks the beginning of protein coding stretches, the start codon. The remaining three triplets

TAG (UAG), TAA (UAA) and TGA (UGA) do not specify amino acids, but any one of them signals the end of a protein coding sequences, a stop codon.

The genetic code is said to be "degenerate" because more than one codon specify the same amino acid. A particular codon never specifies more than one amino acid.

Is the genetic code the same in all organisms? The base sequence of many genes is known as are the amino acid sequences of their encoded proteins. In each case, the nucleotide change in the gene and the amino acid change in the protein are as predicted by the genetic code. Furthermore, m RNAs can be correctly translated by the protein synthesizing machinery of very different species. For example bacteria efficiently express recombinant DNA molecules encoding human proteins such as insulin. These experimental findings strongly suggest that the genetic code is universal with few minor exceptions.

1.8-Translating the open reading frame

Translation is the process of translating the sequence of a messenger RNA (mRNA) molecule to a sequence of amino acids during protein synthesis according to the genetic code.

In the cell cytoplasm, the ribosome reads the sequence of the mRNA in groups of three bases to assemble the protein. Transcription and translation are catalysed by enzymes. The enzymes involved in this process where themselves produced by the process of transcription and translation.

Messenger RNA contains a sequence of bases which read three at a time code for the amino acids used to make protein chains.

Translation begins when the ribosome binds to sequence up stream of start codon (upstream means closer to 5′ end than the start codon). Then the ribosome slides until it reaches the start codon AUG.

Transfer RNA molecules carry amino acids to the mRNA, where the anticodon matches the codon and ensures the placement of the correct amino acid. Amino acids join one at a time onto the growing chain, and a tRNA floats away after it releases its amino acid. Anticodon is a complementary match to a codon. The correct amino acid tRNA recognizes the codon by using its own complementary anticodon.

The ribosome moves along the mRNA, binding a new tRNA molecule and the amino acid it carries. The process continues until a "stop" codon is reached, the polypeptide is complete, and the mRNA is released from the ribosome. rRNA molecules make up part of a ribosome. These molecules help hold ribosomal proteins in place and help locate the beginning of the mRNA message.

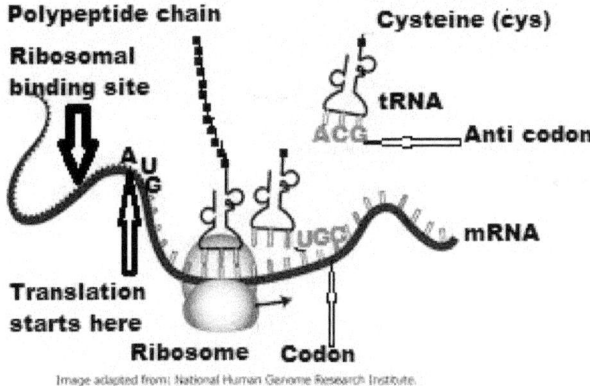

Figure 1.18. Translation of mRNA occurs on the ribosome

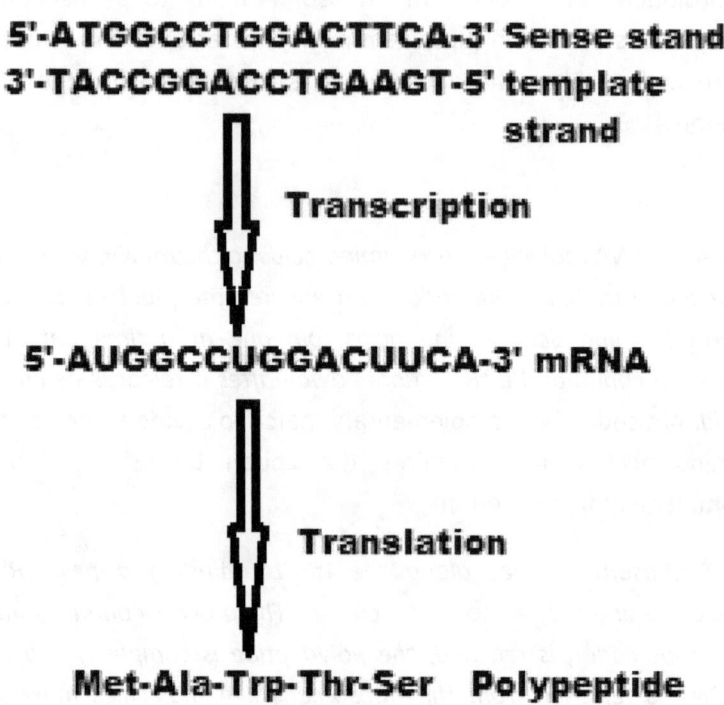

Figure 1.19. Example of the expression of a short DNA sequence

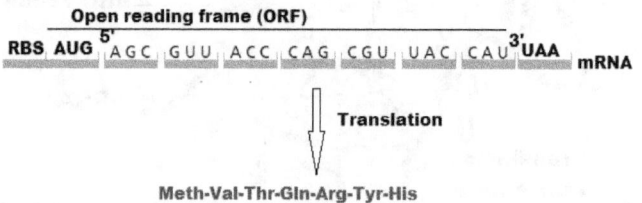

Fig. 1.20. Translating apparatus in the cell recognise and translate the open reading frame.

There are three possible frames on a messenger RNA depending on which of a series of 3 nucleotides is chosen as the first in the first codon. The translation apparatus select the proper frame- the open reading frame by recognizing the start codon-AUG, the AUG codon immediately after the ribosomal binding site is usually the start of translation. Frame is open throughout until a stop codon is reached at the end of the gene's sequence on mRNA.

Figure 1.21. Flow of genetic information from DNA in protein synthesis

1.9-Codon usage bias

Different codons that encode the same amino acid are known as synonymous codons. In a wide variety of organisms, synonymous codons are used with different frequencies. This phenomenon has been termed codon usage bias or codon preference.

The codon usage bias or codon preference is a statistical property of DNA sequences that encode proteins, i.e. open-reading frames.

Natural genes in many species do not use the available codons randomly but show a certain preference for particular codons for the same amino acid.

For example, analysis of genes in *E*.coli shows that some valine codons are used more frequently than others are. The GUU codon is used in 15%, and GUG is used in 38% of the time. This phenomenon of unequal use of codon with identical functions is referred to as codon usage bias.

Another example Argenine is an amino acid that shows very high codon usage bias. Arg can be encoded by 6 different codons, AGG, AGA, CGG, CGA, CGU, and CGC. At random we would expect each codon to be used about 1/6 (17% of the time). In E.coli CGC is used 40% of the time while AGG is used 2 of the time.

Codon usage preference usually correlates with the abundance of t RNAs for a given amino acid, i.e. more frequent codons have more abundant corresponding t RNAs. It is not clear whether codon usage bias drives t RNA evolution or vice versa. The uneven use of codons is too extreme to be accounted for by chance deviation. The genetic code is generally conserved among organisms, but the direction of codon usage bias shifts between different organisms. Thus the identity of the more and less frequent codons for each amino acid may differ between organisms.

For example there are four codons corresponding to the amino acid valine, GUU, GUC, GUA, GUG, all four of the valine therefore we might expect each of the valine codons to be used in about equal proportions. However, this is not the case for many species. For example E.coli prefers the GUG codon for valine 38%, while H.sapiens uses this codon only 10% of the time and instead prefers the GUC codon for valine 40%. The codon usage bias is an important factor in gene expression. This means that generally a gene should use the most frequent codons of the expression host, in order to increase the expression efficiency, for example when trying to express a human gene in E.coli.

1.10-Single point Gene mutations

Mutations: any change in the nucleotide sequence of DNA – genetic information is a mutation. A gene mutation is a change in the order of bases on a strand of DNA. Single point gene mutation refers to a mistake in a single base pair. Mutations cause changes in the genetic code that lead to genetic variation and the potential to develop disease.

Here the possible consequences of mutations of a single base pair in a gene are described.

Single point gene mutations: can be generally categorized into two types.

 Base substitution

Substitution: one base is changed to a different base usually affects no more than a single amino acid and sometimes has no effect at all.

A-Silent mutations. If the DNA sequence CGC is changed to CGA the amino acid arginine will still be produced.

B-Missense mutation, if the codon for arginine CGC is changed to GGC, the amino acid glycine will be produced instead of arginine.

C-Nonsense mutation: alters the nucleotide sequence so that a stop codon is coded for in place of amino acid. The amino acid sequence us cut short and the resulting protein is almost always non-functional.

Base- pair insertions or deletions

Insertion; an extra base is inserted into the DNA sequence, the effect can be dramatic. The grouping of bases shifts in every codon that follows the mutation. Deletions (frame shift mutation) cause a shift in the reading frame of the gene.

Deletion: a base is removed from the DNA sequence, the effects can be dramatic. The groupings of bases shift in every codon that follows the mutation.

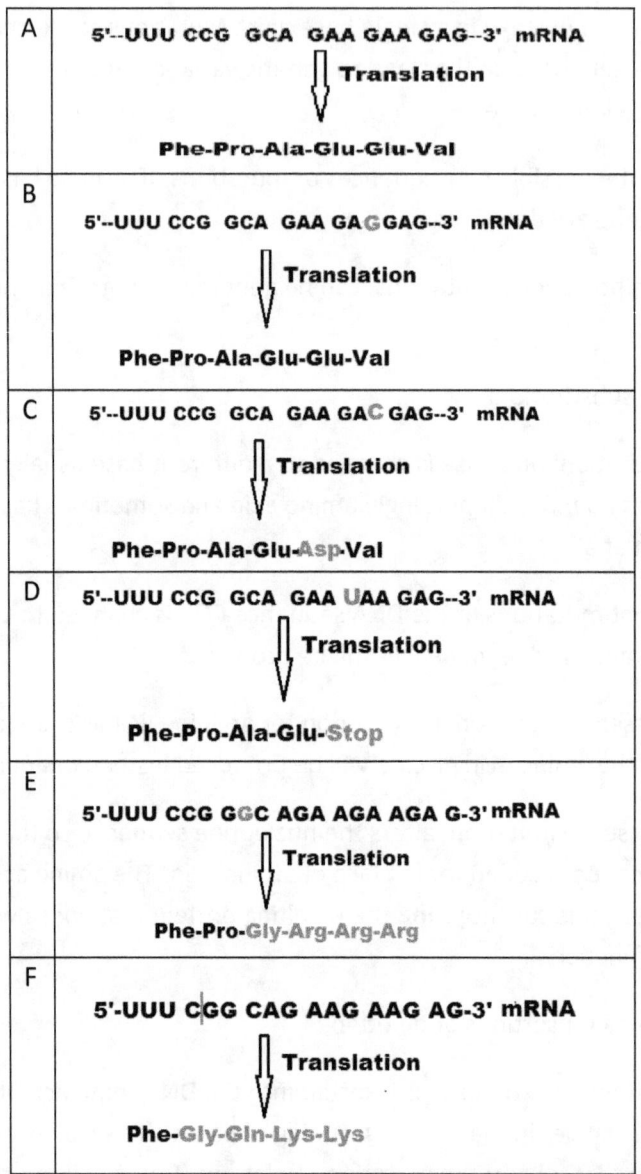

A	5'--UUU CCG GCA GAA GAA GAG--3' mRNA ⬇ Translation Phe-Pro-Ala-Glu-Glu-Val
B	5'--UUU CCG GCA GAA GAG GAG--3' mRNA ⬇ Translation Phe-Pro-Ala-Glu-Glu-Val
C	5'--UUU CCG GCA GAA GAC GAG--3' mRNA ⬇ Translation Phe-Pro-Ala-Glu-Asp-Val
D	5'--UUU CCG GCA GAA UAA GAG--3' mRNA ⬇ Translation Phe-Pro-Ala-Glu-Stop
E	5'-UUU CCG GGC AGA AGA AGA G-3' mRNA ⬇ Translation Phe-Pro-Gly-Arg-Arg-Arg
F	5'-UUU CGG CAG AAG AAG AG-3' mRNA ⬇ Translation Phe-Gly-Gln-Lys-Lys

Table1.1.   shows translation of mRNA in A. Base substitutions in B, C and D. One base insertion in E and one base deletion in F.

1.11-References

1-Tortora, G.J. Microbiology an Introduction 8th, 9th, 10th ed. San Francisco: Pearson
Benjamin Cummings, 2004, 2007, 2010.

2-Biologyfor the IB Diploma Brenda Walpole Ashby Merson-Davies

3-Leighton Dann© Cambridge University Press 2011

4-Biochemistry FOURTH EDITION Reginald H. Garrett • Charles M. Grisham
University of Virginia2010 Brooks/Cole, Cengage Learning

5-*Medical third edition, Cell Biology Edited by Steven R. Goodman, PhD*. Academic Press

2

Control of gene expression in prokaryotes

2.1- Gene regulation in bacteria

Gene regulation is a label for the cellular processes that control the rate and manner of gene expression. A complex set of interactions between genes, RNA molecules, proteins (including transcription factors) and other components of the expression system determine when and where specific genes are activated and the amount of protein or RNA product produced. A prokaryote is a single-celled organism, like bacteria, that doesn't have a nucleus or organelles inside. Most bacteria contain a single, circular chromosome. The bacterial chromosome is usually concentrated in a specific clear region of the cytoplasm called nucleoid. The chromosome is compacted ~1000-fold within the bacterial cell. The principal mechanism of chromosome condensation is through DNA super coiling. In bacteria, approximately half of the DNA super coils are constrained by abundant nucleoid associated proteins (NAPs) that bend or twist the DNA.

The chromosome of bacterium E.coli consists of about 4.6 x 10^6 base pairs. This DNA encodes approximately 4300 genes. The genetic information in the bacterial chromosome is used to encode functions necessary for it to complete its life cycle and its interaction with its environment. Some bacteria might contain also small self-replicating circular pieces of DNA in the cytoplasm called plasmids. Plasmids are non essential extra chromosomal elements that control their own replication.

Some of the gene products are required by the cell under all growth conditions and are called housekeeping genes. These include the genes that encode such proteins as DNA polymerase, RNA polymerase. Housekeeping genes must be expressed at some level all of the time. Frequently, as the cell grows faster, more of the housekeeping gene products are needed. Even under very slow growth, some of each housekeeping gene product is made.

Many other gene products are required under specific growth conditions. These include enzymes that synthesize amino acids, break down specific sugars. The gene products required for specific growth conditions are not needed all of the time. These genes are frequently expressed at extremely low levels or not expressed at all when they are not needed and yet made when they are needed, for example switching genes on and off to make the enzymes needed to digest whatever food sources are available in their growth medium. Control of gene expression is most clearly understood and described in bacteria, and the E.coli is the model experimental organism from which the majority of our knowledge of bacterial transcription has been derived.

Gene expression can be regulated at many of the steps in the pathway from DNA to RNA to protein. The control of transcription is believed to be paramount. This makes sense because only transcriptional control can ensure that no unnecessary intermediates are synthesized. Generally initiation of transcription is the regulated event switching genes on and off in bacteria.

2.2-Operons and regulons

An operon is a group of genes physically linked on the chromosome and under the control of the same promoter. In an operon, the linked genes give rise to a single mRNA that is translated into the different gene products. This type of mRNA is called a polycistronic mRNA.

The most basic and best-understood level gene expression control in bacteria is the regulation of individual operons. Many different types of operon-specific control mechanisms have been described, each of which responds to regulatory signals that are closely related to its function. For example, the expression of an operon encoding genes for a biosynthetic pathway commonly is repressed by pathway-specific end-products such as tryptophan operon, while expression of an operon encoding genes for a catabolic pathway often is activated by pathway-specific substrates such as the lac operon.

The study of operons was the first way that we learned about the regulation of gene expression. Many of the principles of bacterial gene expression were first defined by studies of lactose catabolism in *E. coli,* which can use lactose as its sole carbon source, lactose is a disaccharide sugar found in milk.

Regulons

Regulons are groups of genes that are co-ordinately regulated by nutrient or environmental conditions such as arginine biosynthesis regulon. The genes within a regulon, share a common regulator, usually a protein repressor or activator that recognizes a DNA target sequence common to all members.

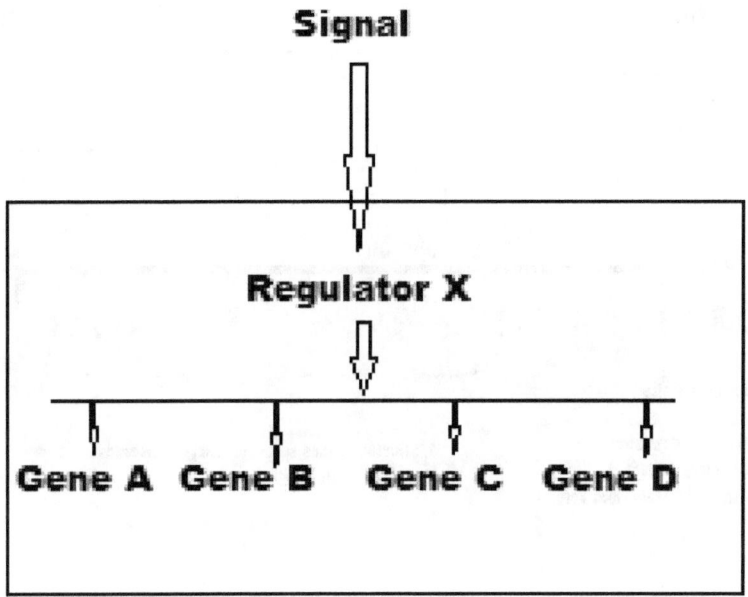

Figure 2.1. A regulon usually have one regulator.

2.3-The lac operon

A good example which illustrates transcriptional control in bacteria is the *E.*coli lac operon. It is called the lac operon because it controls the production of lactose digesting enzymes. The lac operon consist of cluster of three genes that code for the lactose digesting   enzymes.

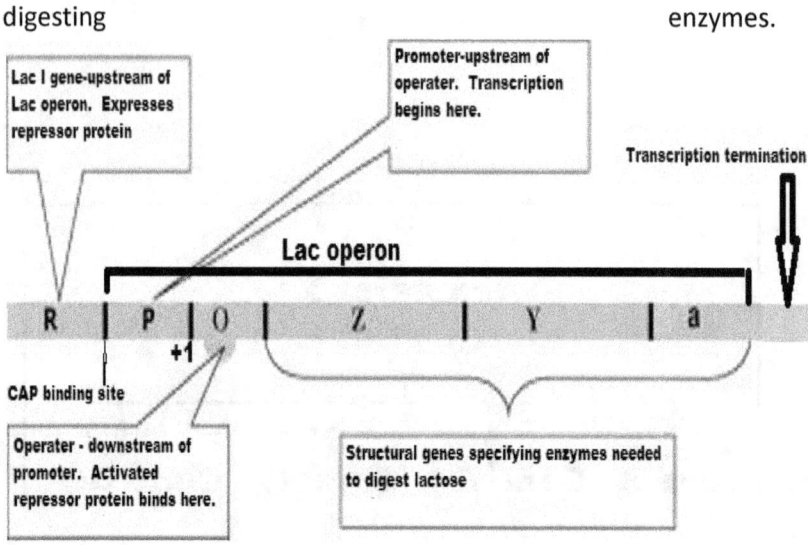

Figure 2.2. Regulation of expression in the lac operon

Main components of the lac operon:

Transcription start

Gene transcription begins at a particular nucleotide shown in the figure as "+1". RNA polymerase actually binds to a site "upstream" (i.e., on the 5' side) of this site and opens the double helix so that transcription of one strand can begin.   Nucleotides before transcription start nucleotide are given negative numbers, and nucleotides after transcription start nucleotide are given positive numbers.

Promoter

The promoter is the region where RNA polymerase binds (at the -10 and -35 regions).   As with the majority of prokaryotic genes, initiation is controlled by two DNA sequence elements that are approximately 35 bases and 10 bases, respectively, upstream of the site of transcriptional initiation and as such are identified as the -35 and -10 positions.

CAP binding site

CAP binding site upstream of lac promoter. The binding of the cAMP-CRP. cAMP receptor protein (CRP) also called catabolite activator protein (CAP) is an example of positive control. This protein binds to an activation site within the promoter region only when it is complexed to cyclic adenosine mono phosphate (cAMP).

Lac I gene

Lac i gene is a regulatory gene with its own promoter that codes for the repressor protein. It happens to be located just upstream of the *lac* promoter. However, it achieves its effect by means of its protein product, which is free to diffuse throughout the cell. The promoter for the lac I gene is always on, but is very week, so it is transcribed only rarely, a gene that is not regulated other than by the strength of its promoter, is said to be constitutive.

Lac repressor

The lac I repressor is the gene product of lac I; It is a protein, with a high affinity for the operator region of the lac operon. (-10 to 0). The repressor binds to the operator (-10 to 0). Lac i protein turns off transcription by binding to the operator and blocking the attachment of RNA polymerase to the promoter.

Operator (O) (repressor site)

Operator is a short sequence of bases that acts like a switch that can be recognized by the repressor protein. The operator controls whether or not transcription will occur, it does this by providing a binding site for the repressor, which blocks RNA polymerase from attaching to the promoter.

Structural genes of the lac operon

Z, y and A are all structural genes (genes that code for polypeptides)

Lac Z is the structural gene coding for beta galactosidase. Beta-galactosidase is an enzyme which converts lactose into the intermediate allolactose and then hydrolyzes this into glucose and galactose.

Lac y for lac permease. Lac permease is an enzyme that transports lactose across the plasma membrane from the culture medium into the interior of the cell

Lac A for transacetylase whose function is still uncertain.

Transcription terminator sequence where RNA polymerase and the newly synthesized RNA dissociate from the DNA to end transcription.

N.B: When a compound is broken down, the process is called catabolic. The lac operon expresses genes whose products catabolise lactose.

Lac operon has 2 regulatory sites. CAP site upstream of lac promoter (activator site) operator site downstream of lac promoter (repressor sit). Glucose and lactose concentrations control the initiation of transcription through their effects on the lac repressor protein and CAP.

Negative control: Absence of lactose

When lactose is absent, the lac i repressor binds the lac operator and shuts off expression of the operon. When the repressor is bound to the operator the RNA polymerase cannot bind to the promoter. No RNA polymerase, no transcription   No enzyme activity.

In the presence of lactose a small amount of it is converted to allolactose (1-6-O-B-galactopyransoyl-D-glucose) by all the few copies of beta gal present in the cell. Allolactose: is a rearranged lactose molecule and an inducer of the lac operon. It binds to lac i repressor and prevents repressor from binding to lac O, leading to transcription of the operon.

Positive control:

cAMP receptor protein (CRP) also called catabolite activator protein (CAP) is an example of positive control. This protein binds to an activation site within the promoter region only when it is complexed to cyclic adenosine mono phosphate (cAMP).

The significance of cAMP is that its concentration is low when intracellular (glucose) is high and vice versa. The other important fact is that cAMP reversibly binds to CAP in a concentration dependent manner.

In summary when intracellular glucose levels are high c AMP is low CAP is not associated with cAMP and is not bound to the DNA. When intracellular (glucose) is low cAMP is high cAMP-CAP complex will increase the frequency of transcription at lac promoter.

Although the presence of lactose removes the repressor, the presence of glucose lowers the level of cAMP in the cell and thus removes CAP.

This arrangement enable the control region of the lac operon to integrate two different signals, so that the operon is highly expressed only when two conditions are met: lactose must be present and glucose must be absent. By having two mechanisms controlling the expression of the lac operon, cells ensure the lac gene products are only made when needed.

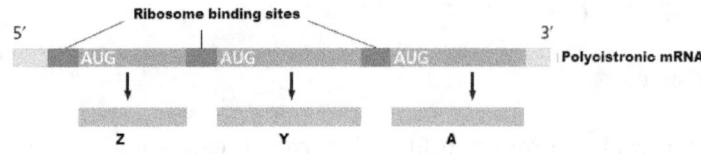

Figure 2.3. Regulation of translation in the lac operon

2.4-Stimulons or Modulons

Certain environmental changes, such as in the osmolarity or oxygen content of the growth medium, may generate several signals that induces genes/operons /regulons. These overlapping networks are referred to as stimulons.

Genes of an operon are physically located together on genome. In contrast, genes of a regulon, a modulon or a stimulon are not necessarily located together.

Gene regulation by a stimulon requires the integration of a variety of nutritional and environmental signals, and the generation of a co-ordinated response that adjusts the basal levels of expression of all genes so as to optimize cell growth and survival under a broad range of possibly rapidly changing conditions.

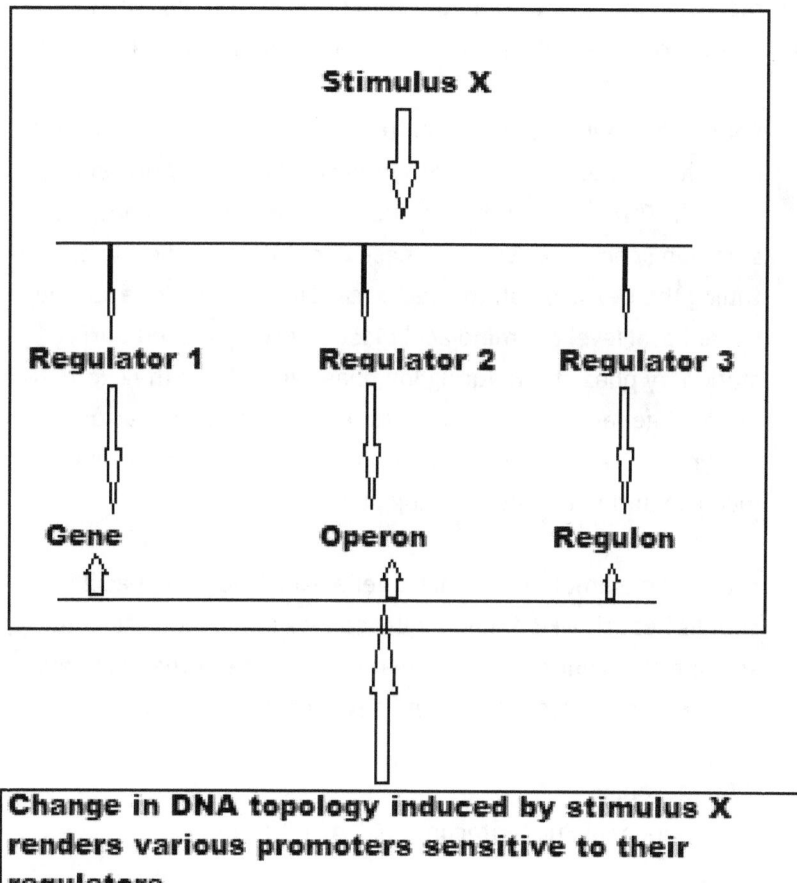

Figure 2.4. Stimulon activation has two co-ordinated effects on DNA topology and generating appropriate stimuli for activation of genes/operons/regulons.

The co-ordinated action of a stimulon is assisted by changes in DNA supercoiling or DNA topology. DNA super coiling is modulated by nutrient and environmental signals. Changes in the

DNA super coiling has an important role in determining the efficiency with which many promoters are transcribed this results in increased sensitivity of promoters to the various new regulators which are produced as a result of activating a stimulon. These two mechanisms allow a stimulon to co ordinate a change in the expression levels of all genes affected by its regulators.

For example when cells are shifted from aerobic to anaerobic growth conditions the rate of synthesis of over fifty proteins is altered. Throughout this transition, all operons, regulons, and stimulon specific controls on these genes remain operative, fine-tuning their expression to specific circumstances. For example, a lower basal level of amino acid biosynthesis is needed during stationary phase than during log phase growth. So the expression levels of genes encoding enzymes required for amino acid biosynthesis are lowered in stationary phase, but must rapidly increase during transition to log phase.

Example of stimulon control on gene expression is in bacillus subtilis heat shock stimulon induced by sudden increase in temp, so it induces genes to create heat shock proteins that will help cope with the stress of the new environment.

2.5-References
1-Adv Biochem Eng Biotechnol (2014) 139: 11–33

2-Annu. Rev. Genet. 2002. 36:175–203

3-The EMBO Journal Vol.17 No.17 pp.4905–4908, 1998

4-Journal of Cell Science 124, 839-845

5-Journal of Cell Science 116, 4067-4075 © 2003

6-Current Opinion in Microbiology 2010, 13:773–780

3

Control of gene expression in complex Eukaryotes

3.1-Gene regulation in complex eukaryotes

Chromatins are the structures that contain the genetic material they are complexes of DNA and proteins. The human genome refers to one complete set of nuclear chromatins, it contains 46 chromatin fibres. The packaging of chromatin is flexible and changes during the eukaryotic cell cycle, inter phase chromatin becomes more tightly packaged at the time of division (mitosis or meiosis), when individual chromosomes become visible as discrete entities.

The human genome contains between 25,000 and 30,000 genes, but there are only 23 pairs of chromosomes. This means each chromosome carries many genes.

Some genes are expressed continuously, as they produce proteins involved in basic metabolic functions; some genes are expressed as part of the process of cell differentiation; and some genes are expressed as a result of cell differentiation. For example Visual pigment genes are expressed only in the cells of the retina. B cells of the pancreas make the protein hormone insulin. Cells of the pancreas make the hormone glucagon. Lymphocytes of the immune system make antibodies. Red blood cells make the oxygen- transport protein haemoglobin.

The process differentiation established a wide variety of cell types (e.g., skin, liver, muscle, etc.); it was not accompanied by any permanent loss of genetic material. Changes in gene expression, rather than losses of genetic material, play a key role in guiding and maintaining cell differentiation.

3.2- Chromatin structure and gene regulation

Storage of eukaryotic DNA in small, compact nuclei requires that this DNA be tightly coiled and compacted in the form of chromatin. However, the structure of chromatin also appears to serve a second, important role, in that it gives eukaryotic cells the capability to exert complex levels of control over gene expression.

The fundamental unit of chromatin is the nucleosome which comprises two copies each of the four core histones, H2A, H2B, H3 and H4. Together they form a histone octamer, which is wrapped inside 147bp of genomic DNA. The DNA bridging two adjacent nucleosomes is termed linker DNA, and is normally bound by the linker histone H1. This primary structure sequence is referred to as beads on string structure as perceived through the electron microscope. Organization of DNA into nucleosomes, constitutes the first stage in the packaging of DNA, subsequently this structure is folded upon itself into much more compact structure, known as the solenoid.

In eukaryotes gene regulation involves long-term changes which allow a cell to become and remain committed to a particular pattern of gene expression. These changes occur prior to a gene becoming active and involve an alternation in the chromatin structure of the whole gene from the tightly packed solenoid structure to the more open "beads on a string" structure. Several lines of evidence support the association of tightly folded compact heterochromatin with gene silencing.

Commitment to a particular pattern of gene expression involves a move from the tightly packed solenoid structure to the more open "beads on a string" structure.

Active genes are believed to be organized into nucleosomes, this appears to be a general characteristic of all transcribed genes and the altered, more open, chromatin structure appears to reflect the ability to be transcribed in a particular tissue or cell type.

Hence, in cells that have become committed to a particular lineage expressing particular genes, such commitment will be reflected in an altered chromatin structure. The altered chromatin structure is believed to arise prior to the onset of transcription and will persist after transcription has ceased.

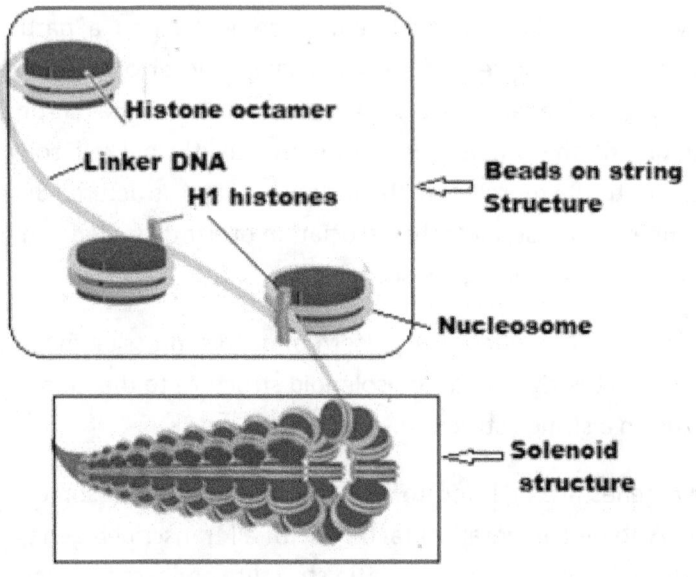

Figure 3.1. Folding of DNA into nucleosomes and solenoid structure

3.3-Epigenetics and gene regulation

Cellular differentiation involves large changes in gene expression concomitant with alterations in genome organisation and chromatin structure. The observed changes in chromatin structure can affect gene expression by changing the accessibility of genes to transcription factors, in either a positive or a negative manner.

Three major classes of such modifications include DNA methylation histone modifications (methylation and/or acetylation) and chromatin remodelling.

1- Histone modifications

Histone proteins are subject a number of modifications and these modifications are known to affect the structure of chromatin. One of the most important histone modifications is histone acetylation, which involves adding acetyl group (COCH3) to the N-terminus of certain lysine amino acids of histone proteins in the nucleosome's histone octamer, thus eliminating the positive charge of affected lysine amino acids.

Histone acetylation is known to result in a more open chromatin structure and these modified histones are found in regions of the chromatin that are transcriptionally active. Conversely, under acetylation of histones is associated with closed chromatin and transcriptional inactivity.

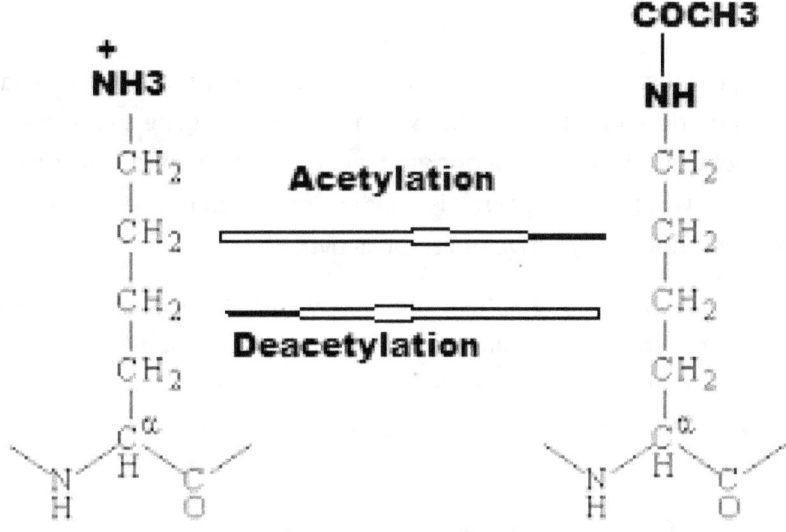

Figure 3.2. Acetylation/deactylation of lysine amino acid

2-DNA methylation

In DNA methylation, a methyl group is added to some cytosine nucleotide. Heterochromatin regions tend to show methylation of specific regions of cytosine bases in specific regions of DNA.

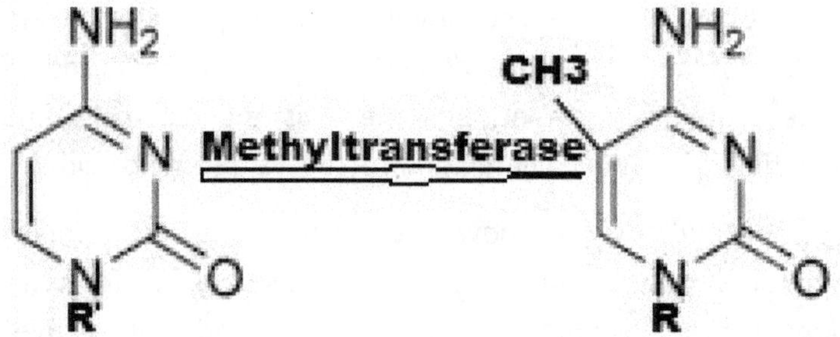

Figure 3.3. DNA methylation of cytosine nucleotide, R: DNA structure

In a general sense what is known about DNA methylation and transcriptional status is that when regions of a gene that can be methylated are, the associated gene is transcriptionally silent and when the region is under methylated the gene is transcriptionally active or can be activated.

When cells undergo differentiation it has been observed that genes that become transcriptionally activated exhibit a reduction in methylation status relative to the level prior to activation and that this under-methylation remains even after transcription ceases.

3- ATP-dependent remodelling of chromatin structure.

ATP-dependent proteins form chromatin associated complexes that remodel the chromatin to allow access of the transcription machinery to the DNA. This consists of mechanisms for moving or displacing histones that depend on the input of energy.

A major role of chromatin remodelling is to change the organisation of nucleosomes at the promoter of a gene that is to be transcribed. This is required to allow the transcription apparatus to gain access to the promoter.

Heterochromatin regions tend to show, reduced acetylation of histone proteins, increased methylation at particular sites on histone proteins and methylation of specific regions of cytosine bases in specific regions of DNA.These molecular changes cause the condensation of the chromatin, which is responsible for its inactivity.

DNA methylation, histone modifications and chromatin remodelling changes are often described as epigenetic because they do not act to alter the primary DNA sequence but instead act at a level just above the DNA sequence. Although DNA methylation, histone modification and chromatin remodelling are not genetic, cells have mechanisms to copy this epigenetic information during their division so that their daughter cells contain the same regulatory data.

Chromatin modifications are usually erased and reset during the production of gametes, such that the adult program of intrinsic cues is replaced with a program more suited to embryonic development.

As a result of variation in chromatin structure in Complex Eukaryotes they have two types of genes in each differentiated tissue; transcriptionally active and transcriptionally silent genes. With regards to bacteria and as a result of variation in DNA topology they also have transcriptionally silent genes which could be activated by sudden and drastic changes in their growth environment which affects DNA topology and renders many promoters more sensitive to the stimulon stimuli resulting from the drastic change in their growth environment.

51

The initiation of transcription in Eukaryotes and prokaryotes are similar in general. Factors which initiate or suppress transcription in Eukaryotes are more complex than prokaryotes.

3.4-References

1-Adv Biochem Eng Biotechnol (2014) 139: 11–33

2-Annu. Rev. Genet. 2002. 36:175–203

3-The EMBO Journal Vol.17 No.17 pp.4905–4908, 1998

4-Journal of Cell Science 124, 839-845

5-Journal of Cell Science 116, 4067-4075 © 2003

6-Current Opinion in Microbiology 2010, 13:773–780

7-Botzman and Margalit Genome Biology 2011, 12:R109

8-Annu. Rev. Genet. 2002. 36:175–203

9-Genes to Cells (2009) 14, 499–509

10-Gene Regulation A eukaryotic perspective Fifth Edition David S Latchman© 2005 by Taylor & Francis Group

4

Recombinant DNA technology

4.1-Recombinant DNA tools and techniques

Recombinant DNA technology is a fruit of several decades of basic research on DNA, RNA, and viruses. The rapid progress in biotechnology indeed its very existence is a result of a relatively few tools and techniques.

Restriction enzymes

Restriction enzymes also called restriction endonucleases recognise and cleave specific base sequence in double-helical DNA by breaking phosphodiester bond. Most recognition sites are palindromes in that both strands exhibit the same sequence when read in the 5' to 3' direction. In recombinant DNA technology restriction enzymes are used as precise, molecular scissors that cut DNA segments at specific DNA sequence.

Many restriction enzymes with different specificities have been isolated from many types of bacteria and are available commercially. The cuts produced by these restriction enzymes may be staggered producing sticky ends or even resulting in blunt ends.

DNA LIGASE

DNA ligase creates a phosphodiester bond between two DNA ends. It forms a phosphodiester bond between 5'phosphate of the donor and 3'hydroxyl of the acceptor. Ligases are found in all organism but the ligaes used in the lab were isolated from bacteria. In recombinant DNA technology commercially available DNA ligases such as T4 DNA ligase which is purified from E.coli cells infected with bacteriophage T4. The availability of many kinds of restriction enzymes and DNA ligases makes it possible to cut DNA sequences from one source and ligate it to another DNA molecule from another source, thus forming a recombinant DNA molecule (figure 2.1).

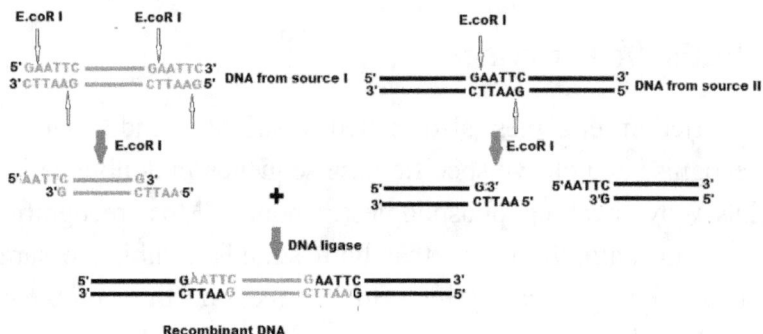

Figure 4.1. Recombinant DNA from two sources.

Agarose gel electrophoresis

Gel electrophoresis is an important laboratory procedure that uses an electrical current to separate Nucleic acids and proteins according to their size. Electrophoresis is the movement of charged particles in an electric field and agarose is a polysaccharide that can be used to form a gel to separate molecule based on size. The method relies on the fact that nucleic acids are poly anionic at neutral pH that is they carry multiple negative charges because of the phosphate groups on the phosphodiester backbone of the nucleic acid strands. Therefore Nucleic acid molecules will move toward the positive electrode of the circuit during gel electrophoresis. The agarose gel contains interlinked agarose molecules which form microscopic pores through which DNA can move. In many types of gels, the electophoretic mobility of a DNA fragment is inversely proportional to the logarithm of the number of base pairs up to a certain limit.

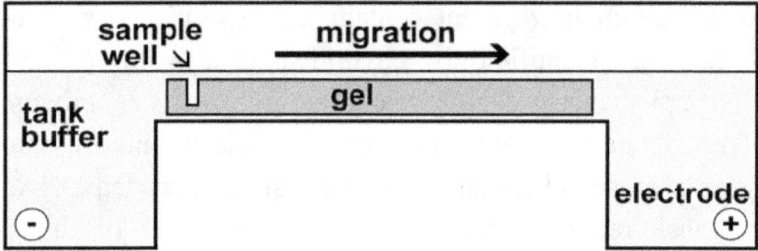

Figure 4.2. Agarose gel electrophoresis cell

Agarose gel can be stained with ethidium bromide, which fluoresces an intense orange when bound to double-helical DNA molecule. A band containing only 50 ng of DNA can be seen. Gel electrophoresis represents the main way by which DNA fragments may be visualized directly.

Detection of specific DNA sequences

DNA probes are oligonucleotides of a known sequence used to probe, or find, DNA to which it is complementary, with whatever sequence the probe binds to, the DNA sequence can be determined. DNA probes are based on the fact that a denatured (heated or chemically treated to become single stranded DNA molecule will hybridise (bind) to sequences that match or are similar to it (figure 2.3). For example synthesise a short 14-20 nucleotides based on the protein sequence using the genetic code, use it as probe. A southern blot allows the detection of a gene of interest by probing DNA fragments that have been separated by electrophoresis with a labelled probe.

A sequence-specific probe is an extremely sensitive detection method, capable of picking out specific DNA or RNA sequences from complex mixtures.

A restriction fragment containing a specific base sequence can be identified by hybridising it with a labelled complementary DNA strand as follows. A mixture of restriction fragments is separated by electrophoresis through an agarose gel, denatured to form single-stranded DNA, and transferred to nitrocellulose sheet. Positions of the DNA fragments in the gel are preserved on the nitrocellulose sheet, where they are exposed to a ^{32}P-labeled single stranded DNA probe. The probe hybridises with a restriction fragment having a complementary sequence and autoradiography then reveals the position of the restriction fragment-probe duplex. Northern plot, probe RNA on a gel with a DNA probe, while a western blot probe a specific protein on a polyacrylamide gel (gel made up of polyacrylamide polymers) using antibody specific for the protein as a probe.

Polymerase chain reaction (PCR)

Polymerase chain reaction (PCR) is a laboratory procedure that can be used to amplify DNA *in vitro*. It is basically an artificial version of DNA replication. In PCR chemically synthesised primers determine where the replication begins compared to origin of replication *in vivo*, two primers for the two strands which allows the DNA synthesis in 5 to 3 direction for both strand (there is no lagging strand) heat is used to separate DNA strands compared to DNA helicase *in vivo*, and a heat resistant DNA polymerase is used. Generally primers of 20 nucleotides or greater will provide a satisfactory sequence. Primers form the initial segment of a polymer that is to be extended on which elongation depends. One of the major factors in the success of PCR is the use of heat resistant DNA polymerase isolated from thermophilic organisms such as Taq polymerase and exhibit an optimal temperature for activity at 72-74°C. Also the temperature necessary for the separation of DNA strands around 94°C will destroy the DNA polymerase activity from other sources.

PCR; DNA synthesis in a tube requires

1-Template DNA (gene you want to copy)

2-Primers specific for the template

3-Free nucleotides (dNTPs)+Mg^{+2}

4-DNA polymerase.

The thermostability of the DNA polymerase makes it feasible to carry out PCR in a closed container, no reagents are added after the first cycle. The product of PCR is a dsDNA molecule that is defined by the 5'ends of the primers. In other words, the length of the DNA molecule is determined by the distance between the primers. One cycle is made up of three steps Denaturation, Annealing and Extension/elongation

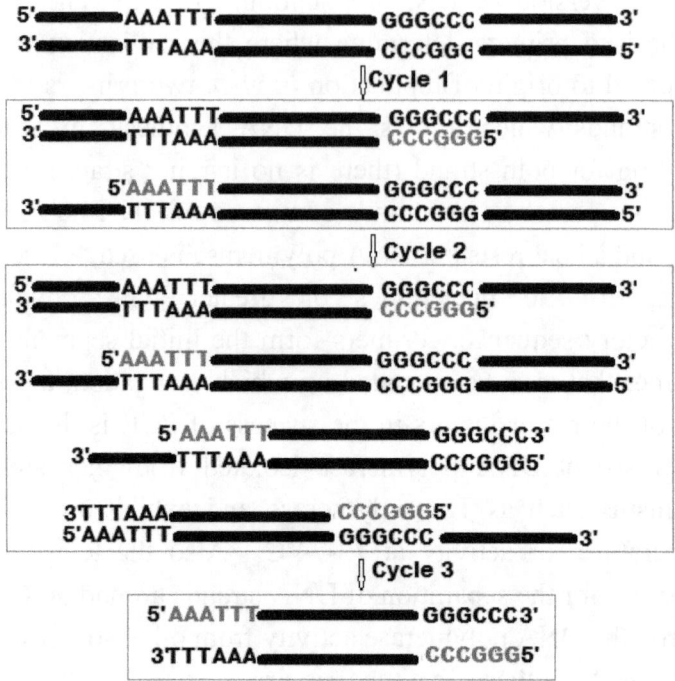

Figure 4.3. Target sequence appears after 3rd cycle, other sequences are present (not shown). Target sequence follows geometric increase in numbers after third cycle. Other components increase to smaller extent and eventually lead to increase in the number of target sequences.

Smaller DNA fragments are amplified more efficiently than longer fragments and it is often particularly difficult to amplify fragments larger than one Kbp. The exact parameters of the PCR reaction needs to optimised for the specific reaction at hand such as heating time, elongation time, time for annealing. At the end of the PCR synthetic, DNA can be sequenced and linkers can be added to this double-helical DNA for ligation to a suitable DNA carrier.

DNA SEQUENCING

DNA sequencing is the determination of the base sequence of a DNA molecule; it is the ultimate analysis of DNA. The DNA sequencing method, developed by Frederick Sanger, is known as dideoxy sequencing. Dideoxynucleotide is a nucleotide that is missing the 3'-hydroxyl group of its sugar. The basis of DNA sequencing in this method is the controlled termination of replication. Sanger method is similar to PCR, except that this method uses only one primer which provides starting reference point, in order to perform sanger sequencing procedure, four test tubes each containing the DNA template to be sequenced, a primer and four deoxynucleotide triphosphates, a DNA polymerase, and each test tube contains one dideoxynucleotide triphosphate. In each test tube the replication of the DNA proceeds until the incorporation of the dideoxynulceotide, which terminates the replication. Results from four test tubes are used to read the sequence of the DNA, the process is now automated.

Reverse transcription

Reverse transcription refers to the synthesis of DNA sequence from RNA template using reverse transcriptase enzyme, which is the reverse of transcribing RNA from DNA template. This enzyme occurs naturally in retroviruses such as avian myeloblastosis virus (AMV). Reverse transcriptase is produced when a retrovirus invades a host cell. It enables the virus to reverse transcribe RNA into a single strand of DNA, using nucleotides from the host cell. The new complementary DNA (cDNA) is then converted to double-straned DNA by the enzyme DNA polymerase. The original RNA is digested and the double stranded DNA is inserted into the host chromosome.

The human gene is usually present in the chromosome as gene with regulatory sequences and the coding information is interrupted by non-coding information. Complementary DNA or cDNA is most often synthesized from mature (fully spliced) mRNA from eukaryotic cells using the enzyme reverse transcriptase. The gene of interest used is normally cDNA when inserting into eukaryotes or prokaryotes cannot splice eukaryotic genes, yeast can splice human genes but often splice human genes incorrectly while mammalian cells can splice eukaryotic genes but alternative splicing may occur.

The purified cDNA is then treated with a restriction enzyme in order to generate fragments with ends capable of being linked to those of the vector (A vector is a DNA molecule used as a vehicle to artificially carry foreign genetic material into another cell). If necessary, short double-stranded segments of DNA (linkers) containing desired restriction sites may be added to create end structures that are compatible with the vector. At this stage the aim is to have the open reading frame of the gene with compatible ends for ligation into a suitable vector. An open reading frame (ORF) is the series of codons in the final mRNA that will result in the translation of a protein, from the initiator AUG to the STOP codon. The ORF therefore does not constitute either the entire mRNA or the entire gene.

4.2-Plasmid vectors

A vector is a DNA molecule used as a vehicle to artificially carry foreign genetic material into another cell, where it can be replicated and/or expressed. A vector containing foreign DNA is termed recombinant DNA. Vectors must either have origin of replication compatible with the host which means it can replicate independently of the host, by using existing replication machinery of host organism or the vector contains DNA sequences that can drive integration into the host genome. Origin of replication is required because the host cell enzyme DNA polymerase which makes copies of the DNA does not initiate the process at random but at selected sites known as origin replication, without origin of replication DNA could not be replicated independently in the host.

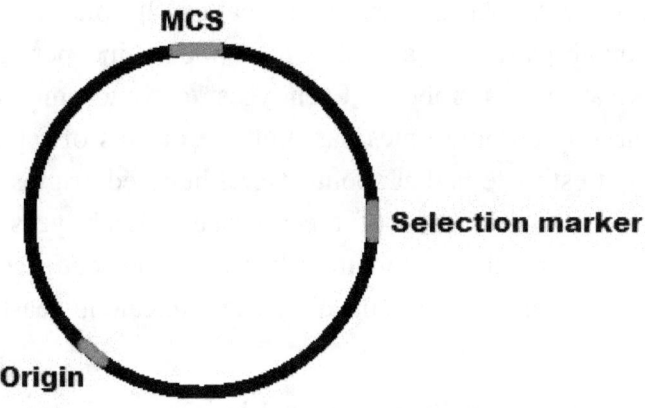

Figure 4.4. Components of plasmid vector.

Plasmids are extra chromosomal double-stranded circular DNA molecule, found widely in many bacteria, for example *E.*coli, but may also be found in a few eukaryotes, for example in yeast such as *Saccharomyces* cerevisiae. Plasmids are the major type of vectors used in recombinant DNA technology in the production of available therapeutic recombinant proteins.

Plasmids in nature have origin of replication compatible with their host cell, thus they are able to replicate independently in their natural host cells. For example the *E*.coli replication origin Oric is approx 240 bp DNA present at the start site for replication of *E*.coli chromosomal DNA. Plasmids containing OriC are capable of independent replication in *E*.coli cells. Sequences located in the vicinity of the origin of replication regulate the copy number of plasmids per cell, these sequences together with origin of replication are called replicons.

Yeasts have plasmid in nature called 2μ plasmids (yeast episomal plasmids) are able to replicate independently in yeast at high number. Each yeast chromosome, like all eukaryotic chromosomes has multiple origins of replication; current estimate is that about several hundred origins exist in the 17 chromosomes of *S*.cerevisiae. Each yeast origin sequence called an autonomously replicating sequence (ARS) confers on any plasmid, the ability to replicate in yeast.

Plasmids are usually represented by simple circles with the important features noted such as origin of replication, antibiotic resistance genes and restriction enzymes sites. In terms of utilisation, plasmids applied as vector (carrier) of DNA can be classified into two groups, cloning and expression vectors.

Plasmid cloning vectors

A cloning vector is carrier DNA plasmid that is designed to amplify a piece of DNA. The DNA polymerase bind to origin of replication, initiating replication of the circular plasmid, once DNA replication is initiated at origin of replication it continues around the circular plasmid regardless of its nucleotide sequence. Thus any DNA sequence inserted into such a plasmid is replicated along with the rest of the plasmid DNA. Bacterial plasmids are usually selected as cloning vectors. In order to be useful as a cloning vector it should have the following features;

1- Convenient restriction enzyme sites for inserting DNA of interest. Some plasmids are artificially modified in order to contain several restriction enzymes sites also called (a multiple cloning site) is the location in a plasmid where a sequence of DNA, typically a gene, can be inserted.

2-Origin of replication; The replication of the plasmid under relaxed control. Which means these plasmids may be present in copies of 10-700 per cell, compared with the tightly controlled copy number of plasmids in nature which mean several copies per host cell.

3-Selectable marker, something that lets you know if the vector is in the host cell in order to identify host cells which carry the plasmid. Example of selectable marker: a drug-resistance gene enzyme that inactivates a specific antibiotic for example the ampicillin gene amp^r encodes B-lactams which inactivates the antibiotic ampicillin.

4-The plasmid should be small and easy to isolate and manipulate

5-When using plasmid as a cloning vector there is usually a restriction on the size of DNA which can be inserted into plasmid. For example Puc18 bacterial plasmids cannot accept DNA molecules larger than 5000 base pairs.

Plasmid expression vectors

Expression vectors are designed to carry DNA in a form suitable for protein expression, Protein expression refers to the way in which proteins are synthesised, modified and regulated in living organism. Naturally occurring plasmids have been drastically altered to serve as expression vectors. Recombinant DNA technology depends on plasmids for protein expression in prokaryotes and eukaryotes. Plasmid expression vectors should carry all the features of the plasmid cloning vector in addition it should carry host signals that facilitate transcription and translation of an inserted gene. The required gene regulatory sequences for transcription and translation should be compatible with the host cell and designed for high expression in the host cell, they normally supplied by the expression vector. Plasmids can be used as expression vectors in bacterial cells, yeasts and mammalian cells. Most of current protein production relies on the use of plasmid expression vectors in one of the three types of cells.

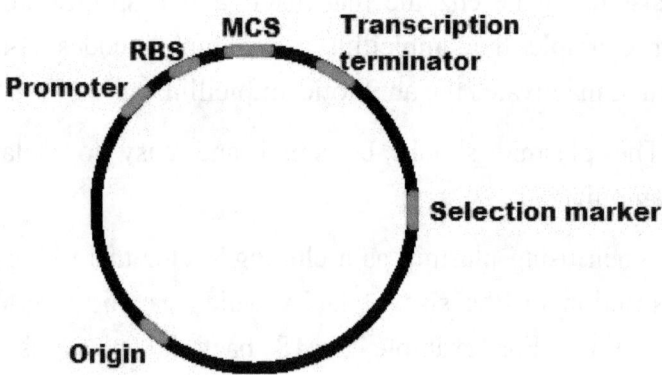

Figure 4.5. Components of plasmid expression vector.

Specific DNA sequence in recombinant expression vector	Function
E.coli origin	Enables the expression vector to replicate independently in the *E.coli* host
Multiple cloning site	Enables inserting the cDNA of the gene
Selection markers	Ideally two selection markers are needed One within the MCS and another outside MCS to select clones which contain the recombinant expression vector
E.coli transcription signals	Strong regulatable promoter upstream (before) MCS and transcription terminator downstream of MCS
E.coli translation signals	RBS upstream of MCS
cDNA containing open read frame of gene	ORF of gene inserted in MCS

Table 4.1. Key components of recombinant plasmid expression vector for expression in *E.coli*.

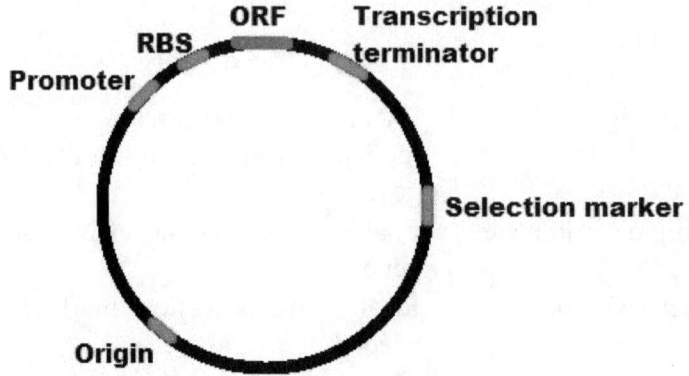

Figure 4.6. Components of recombinant plasmid expression vector.

Plasmid Shuttle expression vectors

Plasmids designed for protein expression in yeast or mammalian cells , are designed to contain all components of *E*.coli cloning vector (*E*.coli origin, selection marker in *E*.coli and MCS) and the required components of expression vector in the eukaryotic host, such expression vectors are called shuttle vectors.   This allows for manipulation such as gene cloning and manipulation required for high expression of the plasmids to be carried out in *E*.coli before transforming yeast or other eukaryotes.

Specific DNA sequences in a shuttle vector	Function
E.coli origin	Enables the expression vector to replicate independently in the *E*.coli host
Multiple cloning site	Enables inserting the cDNA of the gene
Selection markers	Ideally two selection markers are needed One within the MCS and another outside MCS to select clones which contain the recombinant expression vector
Viral origin	Enables the vector to integrate in the host chromosomes.
selection marker	Within the MCS for mammalian expression, enable selection of cells which have integrated the expression vector
Mammalian transcription regulator	Mammalian constitutive promoter for transcription upstream of MCS and transcription terminator downstream of MCS
Mammalian translational signals	RBS and polyadenylation signals

Table 4.2. Components of plasmid shuttle expression vector for cloning in *E*.coli and a mammalian cell.

An expression vector containing a eukaryotic DNA sequence is called recombinant expression vector or expression construct, in order for the eukaryotic DNA to transcribed and translated it needs to be inside a suitable host cell in a process called transformation for bacteria and transfection in eukaryotes. The expression vector and its suitable host cell are called expression system, it is a living cell factory for producing a specific therapeutic protein . Production systems for recombinant proteins include bacteria, yeast, insect cells, mammalian cells, and transgenic animals and plants. Most of recombinant therapeutic proteins approved by the FDA or EMA are normally produced by three types of cells *Escherichia* coli, *Sacchromyces* Cerevisiae and CHO cell line.

The main components of a protein expression system

1-cDNA which has the genetic code for the protein of interest, usually called DNA insert.

2-Expression vector which is designed for transcription and translation of the DNA insert.

3-Host cell; is the cell which designed to house the recombinant expression vector, and produce the recombinant proteins.

.

4.3-Expression systems

1-*E*.coli expression system

-*E*.coli is a Gram –ve rod shaped bacterium. *E*.coli was the first host used to express therapeutic protein and is still considered the workhorse in the field.

- *E. coli* is an easy organism to grow: it grows rapidly and in very large numbers in fairly cheap media. The strains used for protein production are not harmful to humans or to the environment.

68

-In *E*.coli recombinant proteins are normally either directed to the cytoplasm, which may form inclusion bodies (protein aggregates) or expression is directed to the periplasm (space between cell wall and cell membrane) of the bacterial cell.

-Proteins directed to the cytoplasm are normally the most efficiently expressed. It is sometimes possible to renature the protein and regains activity. Disulfide bond formation normally occurs in the periplasm.

-Formation of inclusion bodies can be advantageous in that they generally allow greater levels of expression, they can be separated from large proportion of cytoplasmic proteins by centrifugation giving an effective purification techniques. Major disadvantages extraction of the protein of interest generally requires the use of denaturing agent e.g. urea.

-A prominent and very particular chemical constituent of the outer membrane of Gram-ve bacteria is a chemical named lipopolyaccharide (LPS), biopharmaceutical products gained from Gram –ve organisms must be extensively purified, LPS set free during the isolation of the product has, even in a very low concentration, severe toxic effects to man and animals, also called endotoxins or pyrogens.

-*E*.coli has limited capacity to perform post translational modifications, the simplest of these modifications is the removal of the N-terminal methionine residue which can occur in all organisms. E.coli does not perform many posttranslational modifications such as N-linked glycosylation or O-linked glycosylation.

- If the recombinant protein is glycosylated in nature, and the non-glycosylated form achieves the desired therapeutic action, then it is probably can be made by E.coli expression system. For example Filgrastim [recombinant human granulocyte colony stimulating factor (G-CSF)] marketed as (Neupogen®) is made by bacterial cell is not glycosylated while the natural counterpart is glycosylated.

General procedure of constructing *E.coli* expression system

To express a eukaryotic protein in *E. coli,* the eukaryotic cDNA must be cloned in an *expression vector* that contains regulatory signals for both transcription and translation. Accordingly, a *promoter* where RNA polymerase initiates transcription as well as a *ribosome-binding site* to facilitate translation are engineered into the vector just upstream from the restriction site for inserting foreign DNA, in addition the vector should contain a transcription terminator signal downstream of the restriction site. The AUG initiation codon that specifies the first amino acid in the protein (the *translation start site*) and translation stop codon are normally contributed by the cDNA insert.

The expression vector is treated with a restriction endonuclease to cleave the DNA at the site where foreign cDNA will be inserted. The restriction enzyme is chosen to generate a configuration at the cleavage site that is compatible with that at the ends of the foreign cDNA. Typically, this is done by cleaving the vector DNA and foreign cDNA with the same restriction enzyme. To improve efficiency of ligation procedure, the cleaved vector may be treated with an enzyme (alkaline phosphatise) that dephosphorylates the vector ends. Vector molecules with dephosphorylated ends are unable to rejoin during the ligation reaction, which can improve the efficiency of ligation. Recombinant DNA molecules are generated by simply mixing the prepared vector DNA with the prepared foreign cDNA in the presence of DNA ligase.
Transformation

Expression Vectors containing the foreign gene of interest (recombinant expression construct) must be incorporated into living cells so that they can be replicated and expressed. The cells receiving the vector are called host cells, and once they have successfully incorporated the vector they are said to be transformed. Vectors are large molecules which do not readily cross cell membranes, so the membranes must be made permeable in some way. There are different ways of doing this depending on the type of host cell.

Heat Shock. Cells are incubated with the vector in a solution containing calcium ions at 0°C. The temperature is then suddenly raised to about 40°C. This heat shock causes some of the cells to take up the vector, though no one knows why. This works well for bacterial and animal cells.

Electroporation. Cells are subjected to a high-voltage pulse, which temporarily disrupts the membrane and allows the vector to enter the cell. This is the most efficient method of delivering genes to bacterial cells.

Screening

After transformation with the recombinant expression vector the bacteria are plated on antibiotic containing media , if the MCS contains antibiotic resistance gene which is interrupted by inserting the foreign gene. The antibiotic will kill the bacteria which do not contain plasmid. Therefore, all of the colonies will represent bacteria with plasmid.

Clones of the cDNA can be screened on the basis of their capacity to direct the synthesis of a foreign protein in bacteria. A radioactive antibody specific for the protein of interest can be used to identify colonies of bacteria that harbour the corresponding cDNA expression vector. Spots of

71

bacteria on replica plate are lysed to release proteins, which bind to an applied nitrocellulose filter A^{125}I-labeled antibody pecific for the protein of interest is added, and autoradiography reveals the location of the desired colonies on the master plate. This immunochemical screening approach can be used whenever proteins are expressed and corresponding antibody is available.

Clone selection

Select high expression transformants example by antibiotic gradient, then expression evaluation and optimization. The most effective cell line is selected for expansion. During selection, the cells that can produce the recombinant proteins most effectively are identified and expanded to manufacture the medicine. This cell line is unique to each manufacturer and is the source of all future product.

Example of *E.coli* expression system.

A widely used protein expression system is based on the pET plasmid. Transcription of the cloned gene insert is under the control of the bacteriophage T7 RNA polymerase promoter in pET. This promoter is not recognized by the *E.coli* RNA polymerase, so transcription can only occur if the T7 RNA polymerase is present in host cells. Host *E. coli* cells are engineered so that the T7 RNA polymerase gene is inserted in the host chromosome under the control of the *lac* promoter such as The *E. coli* strain BL21 (DE3) is used for over expression of therapeutic recombinant proteins. It has the advantage of being deficient in both *lon* and *ompT* proteases and contains a copy of the T7 RNA polymerase gene under the control of the lac promoter. This modification enable stable expression of protein using T7 promoter driven constructs. IPTG induction triggers T7 RNA polymerase production and subsequent transcription and translation of the pET insert. The bacteriophage T7 RNA polymerase is so active that most of the host cell's resources are directed into protein expression.

Small scale test expression

Small-scale test expression is widely used as a predictive tool to determine which of the derivative clones actually produces soluble protein and to establish the optimal scale for the large-scale growth. Simplest method is to prepare crude extract of total protein from the expressing cells at different times after induction of expression and analyze them by SDS-PAGE the gel should show clear evidence of new protein being synthesized after induction.

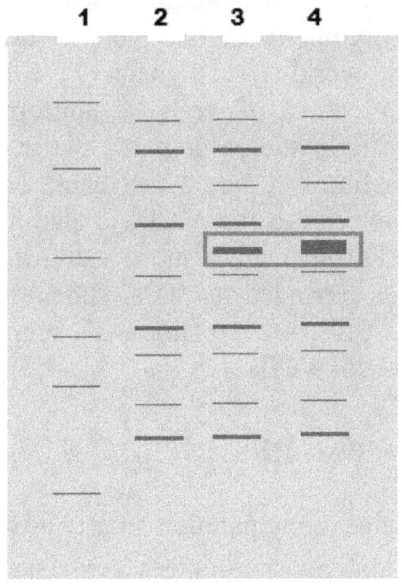

Figure 4.7. SDS-PAGE; lane1: molecular weight markers, lane 2: bacterial cell lysate before induction of expression, lanes 3 and 4, bacterial cell lysate after different times of expression induction.

After evaluation of key parameters, a single production clone is chosen and banked as frozen vials (cell banks) for future use. The chosen clone is subsequently used for generating clinical trial materials and if successful eventually employed in commercial manufacturing.

Cell Banking

Cell banks are used to reduce the variability of the starting material, which is usually the greatest source of variability for biological products. Cell banking assures that an adequate supply of equivalent, well-characterized cells exist for production of recombinant protein. In addition to providing a constant supply of biological starting material. The cell bank system normally consists of two tiers: a master cell bank (MCB); and a working cell bank (WCB). The MCB represents a collection of cells of uniform composition derived from a single source prepared under defined culture conditions. Sometimes, a parent cell bank, comprising vials of the progenitor cells to the MCB, is also maintained. The WCB is derived from one or more vials of cells from the MCB, which are expanded by serial subculture. Generally, the production cells are obtained by expansion of the Working Cell Bank (WCB).

2-Yeast expression systems

-Yeast species have many parallels to *E*.coli when it comes to protein production. Saccharomyces cerevisiae; the common baker's yeast is the most widely used for protein production.

-*S*.cerevisiae regulator advantage it is regarded as safe and so there are fewer hurdles to overcome when trying to establish the safety of new product purified from it.

-Yeasts are genetically easy to manipulate, can be grown quickly and cheaply high cell densities, as eukaryotes they are better suited to the production of eukaryotic proteins.

-Some yeast strains are very efficient at producing large amounts of protein and secreting it to the growth media, which simplifies the harvesting and purification of proteins.

-Non-glycosylated protein are typically expressed in either *E*.coli or the yeast *Sachromyces* cerevisiae, some a glycosylated proteins are expressed in both systems.

Shuttle vectors designed for protein expression in yeast are based on the yeast 2μ plasmid. These vectors have essential components for gene cloning in *E*.coli and the components for gene expression in yeasts. The yeast shuttle vector needs multiple origins of replication, multiple selection markers, different promoters etc. Selection marker in yeast is not antibiotic resistance, Example a host strain of *S*.cerevisiae with mutation in the trpI gene will be unable to grow in medium which does not contain tryptophan. If the vector plasmid carries tryptophan gene then selection can be performed on media lacking tryptophan.

Example of recombianat protein produced by *S*.cerevisiae is Granulocyte-maceophage colony stimulating factor (GM-CSF) Leukine, is a single chain glycosylated polypeptide of 127 amino acid residues expressed from *S*.cerevisiae. The glycoprotein consists of three molecular species having relative molecular weights of about 19500, 16800 and 15500 due to different levels of glycosylation.

3-Mammalian cell-based expression

Cells from rodent species have the interesting property that when grown in culture, most eventually die but a small proportion continue to grow indefinitely: these are referred to as being immortalized and are widely used for production of recombinant proteins. Several different "lines" of cells are used, which have been established from different rodent tissues, the commonest being a cell line established from the ovary of a Chinese hamster, called CHO cells.

This is the most expensive approach to recombinant protein production, especially as the possible co-purification of viruses with the protein means that rigorous quality control procedures must be employed to ensure that the product is safe.

Many of the proteins with pharmaceutical uses are large, complex proteins with many disulfide bonds and with glycans attached. Neither *E. coli* nor yeast is yet able to produce all these proteins in a state suitable for use for many glycoproteins. So despite the fact that there are some drawbacks with using cultured mammalian cells, this remains a highly important method for producing recombinant proteins.

Mammalian expression systems are used primarily to generate secreted rather than intracellular recombinant proteins, serum-free media have been developed which simplifies the purification of secreted recombinant proteins. Cell culture conditions such as nutrient content, ph, temperature, oxygen levels and ammonia can significantly affect the glycosylation profile.

Mammalian cells do not have plasmids in nature, but appropriately modified bacterial plasmid, normally *E*.coli plasmid in the lab can be introduced into a mammalian cell where it could integrate into the chromosomes of mammalian cells under the appropriate conditions, which are hence passed to daughter cells on cell division in the same way as any other gene.

Shuttle vectors designed for protein expression in mammalian cells such as CHO, are usually *E.*coli plasmids with essential features of plasmid cloning vector, in addition they should contain sequences which allow integration of the DNA and surrounding regulatory sequences for expression in mammalian cells.

Expression vectors and cloned genes are still used to maximize yields, by placing the gene under control of a promoter that is stronger than the one to which it is normally attached. This promoter is often obtained from viruses such as SV40, cytomegalovirus (CMV), or Rous sarcoma virus (RSV). In addition they should also contain DNA sequences for selection marker. Example; Dihydrofolate reductase (DHFR) selection cassette along with an expression cassette containing the gene of interest. DHFR converts dihydrofolate into tetrahydrofolate which is required for the synthesis of glycine, hypoxanthine (a purine derivative) and thymidine which are essential compounds for cell survival. Methotrexate which binds and inhibit DHFR is used as amplification agent for DHFR along with the gene of interest. An example of plasmid pcDNA 3.3 plasmid, expression is driven by the immediate early promoter of human cytomegalovirus (CVM). This is a strong promoter, constitutively active in mammalian cells.

The standard method for stable CHO expression involves transfecting dihydrofolate reductase (DHFR) deficient CHO cells such as DXB11 DG44 with a DHFR selection cassette along with an expression cassette containing the gene of interest. Several methods like calcium phosphate precipitation and electroporation are commonly used with optimised protocols available in the literature. Clonal selection is then performed by growth in media devoid of glycine, hypoxanthine and thymidine. The DHFR system also enables efficient amplification of the cloned DNA. When cultured in high levels of methotrexate (MTX), a folic acid analogue that blocks DHFR activity, transfected cells must cope with the decrease in DHFR activity.

Frequently, a particular therapeutic protein is available from two different expression systems. Insulin and human growth hormone, for example, are expressed either in *E. coli* or in *S. cerevisiae*.

4.4-References

1-Garret, R.H., Grisham, C.M.,2010, Biochemistry 4th edition, Brooks/Cole.

2-Watson, J. D., Gilman, M., Witkowski, J., and Zoller, M., 1992. *Recombinant DNA* (2d ed.). Scientific American Books.

3-Ausubel, F. M., Brent, R., Kingston, R. E., and Moore, D. D., (Eds.) 1999. *Short Protocols in Molecular Biology: A Compendium of Methods from Current Protocols in Molecular Biology.* Wiley.

4-Nicholl, D.S.T., 2008, An introduction to genetic engineering, 3rd edition. Cambridge University Press.

5-Dale, J.W., Schantz, M.V., 2002, From Genes to Genomes: Concept and applications of DNA technology. John Wiley and sons Ltd.

6-Brown,T.A., 2010, Gene cloning and DNA analysis: An introduction, sixth edition. Wiley-Blackwell.

7-Lodge, J., Lund, P., Michin,S., 2007,Gene cloning: Principles and applications. Taylor and Frances Group.

8-Schumann, W. and Ferreira, L. 2004. Production of recombinant protein in *Escherichia coli*. *Genetics and Molecular Biology*. **27**, 442–453.

9-Sodoyer, R. 2004. Expression systems for the production of recombinant pharmaceuticals. *Biodrugs* **18**, 51–62.

10-Brown, T. 2006. *Gene Cloning and DNA Analysis, an Introduction*. Blackwell Science, UK.

Gellissen, G (ed.). 2005. *Production of Recombinant Proteins*. Wiley-VCH, Germany.

11-Butler, M. 1996. *Animal Cell Culture and Technology. The basics*. IRL Press.

12-Merten, O.-W., Mattanovich, D., Lang, C., Larsson, G., Neubauer, P., Porro, D., Postma, P., Teixeira de Mattos, J.,and Cole, J.A. *Recombinant Protein Production with Prokaryotic and Eukaryotic Cells: A Comparative View on Host Physiology*? Kluwer.

5-Glossary

Adenine (A) Nitrogenous base found in DNA and RNA.

Affinity chromatography Technique used to separate cellular components by selective binding and release using an appropriate medium to which the component of interest will bind. An example is the selection of mRNA from total RNA using oligo(dT)-cellulose.

Agarose Jelly-like matrix, extracted from seaweed, used as a support in the

separation of nucleic acids by gel electrophoresis.

Ampicillin (Ap) A semisynthetic ß-lactam antibiotic.

Annealing Attachment of an oligonucleotide to a single-stranded DNA molecule by hybridization.

Antibody An immunoglobulin that specifically recognises and binds to an antigenic determinant on an antigen.

Anticodon The three bases on a tRNA molecule that are complementary to the codon on the mRNA.

Antiparallel The arrangement of complementary DNA strands, which run in different directions with respect to their 5_→3_ polarity.

Antisense RNA Produced from a gene sequence inserted in the opposite orientation, so that the transcript is complementary to the normal mRNA and can therefore bind to it and prevent translation.

Autoradiograph Image produced on X-ray film in response to the emission of radioactive particles.

Auxotroph A cell that requires nutritional supplements for growth.

Bacteriophage A bacterial virus.

Beta galactosidase An enzyme encoded by the *lacZ* gene. Splits lactose into glucose and galactose.

Biotechnology Generic word to describe the application of bioscience for the benefit of humankind. Encompasses a wide range of disciplines and procedures but is often mistakenly thought to refer exclusively to the industrial scale use of genetically modified microorganisms.

Blunt ends DNA termini without overhanging 3' or 5'ends. Also known as *flush ends*.

Cap A chemical modification that is added to the 5'end of a eukaryotic mRNA molecule during post-transcriptional processing of the primary transcript.

Capsid The protein coat of a virus.

Carbohydrate Molecule containing carbon, hydrogen, and oxygen, empirical formula CH_2O. Important in energy storage and conversion reactions in the cell; examples include glucose, fructose, and lactose. Polymers of carbohydrates are known as polysaccharides and are used as energy storage compounds.

Cassette A DNA sequence consisting of promoter–ribosome binding site–unique restriction site–terminator (or for a eukaryotic host, promoter–unique restriction site–polyadenylation sequence) carried by certain types of expression vector. A foreign gene inserted into the unique restriction site is placed under control of the expression signals.

Cell bank Vials of cells of uniform composition (although not necessarily clonal) derived from a single tissue or cell, aliquoted into appropriate storage containers, and stored under appropriate conditions, such as the vapor phase of liquid nitrogen.

Cell extract A preparation consisting of a large number of broken cells and their released contents.

Cell line Cells that have been propagated in culture since establishment of a primary culture and survival through crisis and senescence. Such surviving cells are immortal and will not senesce. Diploid cell strains have been established from primary cultures and expanded into cell banks but have not passed through crisis and are not immortal. [The ATCC uses the abbreviation CCL to signify their Certified Cell Lines.]

cDNA DNA that is made by copying mRNA using the enzyme reverse transcriptase.

cDNA library A collection of clones prepared from the mRNA of a given cell or tissue type, representing the genetic information expressed by such cells.

Cell membrane Lipid bilayer--based structure containing proteins embedded in or on the membrane. Acts as a selectively permeable barrier that separates the cell from its environment.

Cell wall Found in bacteria, fungi, algae, and plants, the cell wall is a rigid structure that encloses the cell membrane and contents. Composed of a variety of polysaccharide-based components such as peptidoglycan (bacteria), chitin (fungi), and cellulose (plants, algae, and fungi).

Chromatography Method of separating various types of molecules based on their affinity or physical behaviour when passed through a matrix and eluted with a suitable solvent.

Chromosome A DNA molecule carrying a set of genes. There may be a single chromosome, as in bacteria, or multiple chromosomes, as in eukaryotic organisms.

Cleared lysate A cell extract that has been centrifuged to remove cell debris, subcellular particles and possibly chromosomal DNA.

Clone (1) A colony of identical organisms; often used to describe a cell carrying a recombinant DNA fragment. (2) Used as a verb to describe the generation of recombinants. (3) A complex organism (*e.g.* sheep) generated from a totipotent cell nucleus by nuclear transfer into an enucleated ovum.

Codon The three bases in mRNA that specify a particular amino acid during translation.

Codon bias The fact that not all codons are used equally frequently in the genes of a particular organism.

Cohesive ends Those ends (termini) of DNA molecules that have short complementary sequences that can stick together to join two DNA molecules. Often generated by restriction enzymes.

Competent Refers to bacterial cells that are able to take up exogenous DNA.

Complementary Two polynucleotides that can base pair to form a double-stranded molecule.

Complementary DNA (cDNA) cloning A cloning technique involving conversion of purified mRNA to DNA before insertion into a vector.

Conformation The spatial organization of a molecule. Linear and circular are two possible conformations of a polynucleotide.

Contaminants Any adventitiously introduced materials (e.g., chemical, biochemical, or microbial species) not intended to be part of the manufacturing process of the drug substance or drug product.

Copy number The number of plasmid molecules in a bacterial cell. (2) The number of copies of a gene in the genome of an organism.

Covalent bond Relatively strong molecular bond in which the electron configurations of the constituent atoms is satisfied by sharing electrons.

C terminus Carboxyl terminus, defined by the COOH group of an amino acid or protein.

Cyanogen bromide Chemical used to cleave a fusion protein product from the N-terminal vector-encoded sequence after synthesis.

Degradation Products Molecular variants resulting from changes in the desired product or product-related substances brought about over time and/or by the action of, e.g., light, temperature, pH, water, or by reaction with an excipient and/or the immediate container/closure system. Such changes may occur as a result of manufacture and/or storage (e.g., deamidation, oxidation, aggregation, proteolysis). Degradation products may be either product-related substances, or product-related impurities.

Deletion Change to the genetic material caused by removal of part of the sequence of bases in DNA.

Deoxynucleoside triphosphate (dNTP) Triphosphorylated ('high-energy') precursor required for synthesis of DNA, where N refers to one of the four bases (A, G, T, or C).

Deoxyribonucleic acid (DNA) A condensation heteropolymer composed of nucleotides. DNA is the primary genetic material in all organisms apart from some RNA viruses. Usually double-stranded.

Deoxyribose The sugar found in DNA.

Deoxyribonuclease (DNase) A nuclease enzyme that hydrolyses (degrades) single- and double-stranded DNA.

Diabetes mellitus (DM) Condition where high levels of blood glucose exist because of problems in regulation of glucose levels. May be caused by insulin deficiency or other defects in the glucose regulation system.

Dideoxynucleoside triphosphate (ddNTP) A modified formof dNTP used as a chain terminator in DNA sequencing.

Diploid Having two sets of chromosomes.

DNA ligase Enzyme used for joining DNA molecules by the formation of a phosphodiester bond between a 5'phosphate and a 3'OH group.

DNA polymerase An enzyme that synthesises a copy of a DNA template.

DNA sequencing Determination of the order of nucleotides in a DNA molecule.

END-OF-PRODUCTION CELLS (EOPC) Cells cultured (under conditions comparable to those used in production) from the MCB or WCB to a passage level or population doubling level comparable to or beyond the highest level reached in production. .

Electrophoresis Separation of molecules on the basis of their charge-to-mass ratio.

Electroporation Technique for introducing DNA into cells by giving a transient electric pulse

Elution The unbinding of a molecule from a chromatography column.

Endonuclease An enzyme that cuts within a nucleic acid molecule, as opposed to an exonuclease, which digests DNA from one or both ends.

Enzyme A protein that catalyses a specific reaction.

Enzyme-linked immunosorbent assay (ELISA) Technique for detection of specific antigens by using an antibody linked to an enzyme that generates a coloured product. The antigens are fixed onto a surface (usually a 96-well plastic plate) and thus large numbers of samples can be screened at the same time.

Escherichia coli The most commonly used bacterium in molecular biology.

Ethidium bromide A molecule that binds to DNA and fluoresces when viewed
under ultraviolet light. Used as a stain for DNA.

Eukaryotic The property of having a membrane-bound nucleus.

Exon Region of a eukaryotic gene that is expressed *via* mRNA.

Exonuclease An enzyme that digests a nucleic acid molecule from one or both ends.

Expression vector A cloning vector designed so that a foreign gene inserted into the vector is expressed in the host organism.

Fermenter A vessel used for the large scale culture of microorganisms.

Extrachromosomal element A DNA molecule that is not part of the host cell chromosome.

Ex vivo Outside the body. Usually used to describe gene therapy procedure in which the manipulations are performed outside the body, and the altered cells returned after processing.

Fibrin Insoluble protein involved in blood clot formation.

Fibrinogen Precursor of fibrin, converted to fibrin by the action of thrombin.

Filtration Method of separating soild and liquid components of a suspension by passing through a filter.

Finished sequence Refers to a completed DNA or protein sequence in which anomalies and missing regions have been resolved.

Flanking Control Regions Non-coding nucleotide sequences that are adjacent to the 5' and 3' end of the coding sequence of the product which contain important elements that affect the transcription, translation, or stability of the coding sequence. These regions include, e.g. promoter, enhancer, and splicing sequences and do not include origins of replication and antibiotic resistance genes.

Food and Drug Administration (FDA) Regulatory body in the USA responsible for approval of medicines and foodstuffs.

Formulation Used to describe the 'recipe'used for the production of a pharmaceutical.

Gel electrophoresis Technique for separating nucleic acid molecules on the basis of their movement through a gel matrix under the influence of an electric field. See *Agarose* and *Polyacrylamide*.

Gene The unit of inheritance, located on a chromosome. In molecular terms, usually taken to mean a region of DNA that encodes one function. Broadly, therefore, one gene encodes one protein.

Gene cloning The isolation of individual genes by generating recombinant DNA molecules, which are then propagated in a host cell that produces a clone that contains a single fragment of the target DNA.

Gene therapy The use of cloned genes in the treatment of genetically derived malfunctions. May be delivered *in vivo* or *ex vivo*. May be offered as gene addition or gene replacement versions.

Genetic code The triplet codons that determine the types of amino acid that are inserted into a polypeptide during translation. There are 61 codons for 20 amino acids (plus 3 stop codons), and the code is therefore referred to as *degenerate*.

Genetic engineering The use of experimental techniques to produce DNA molecules containing new genes or new combinations of genes.

Genome Used to describe the complete genetic complement of a virus, cell, or organism.

Guanine (G) Nitrogenous base found in DNA and RNA.

Guanosine Nucleoside composed of ribose and guanine.

Homopolymer A polymer composed of only one type of monomer, such as polyphenylalanine (protein) or polyadenine nucleic acid.

Host cell A cell used to propagate recombinant DNA molecules.

Hybridisation The joining together of artificially separated nucleic acid molecules *via* hydrogen bonding between complementary bases.

Hydrolysis Reaction where two covalently joined molecules are split apart by the addition of the elements of water. In effect the reversal of a dehydration synthesis reaction.

Impurity Any component present in the drug substance or drug product which is not the desired product, a product-related substance, or excipient including buffer components. It may be either process- or product-related.

Inclusion body A crystalline or paracrystalline deposit within a cell, often containing substantial quantities of insoluble protein.

Induction Of a gene The switching on of the expression of a gene or group of genes in response to a chemical or other stimulus.

In vitro Literally 'in glass', meaning in the test tube, rather than in the cell or organism.

In vivo Literally 'in life', meaning the natural situation, within a cell or organism.

Insulin Protein hormone involved in the regulation of blood glucose levels. Has been available in recombinant formsince the 1980s.

Intervening sequence Region in a eukaryotic gene that is not expressed *via* the processed mRNA.

Intron See *Intervening sequence*.

Integration Site The site where one or more copies of the expression construct is integrated into the host cell genome.

In-vitro Cell Age Measure of time between thaw of the MCB vial(s) to harvest of the production vessel measured by elapsed chronological time in culture, by population doubling level of the cells, or by passage level of the cells when subcultivated by a defined procedure for dilution of the culture.

IPTG *iso*-propyl-thiogalactoside, inducer that de-represses transcription of the *lac* operon.

Kilobase (kb) 10^3 bases or base pairs, used as a unit formeasuring or specifying the length of DNA or RNA molecules.

Ligase (DNA ligase) An enzyme that, in the cell, repairs single-stranded discontinuities
in double-stranded DNA molecules. Purified DNA ligase is used in gene cloning to join DNA molecules together.

Linker A synthetic self-complementary oligonucleotide that contains a restriction enzyme recognition site. Used to add cohesive ends to DNA molecules that have blunt ends.

Macromolecule Large polymeric molecule made up of monomeric units, commonly used to describe proteins (monomers are amino acids) and nucleic acids (monomers are nucleotides).

Manufacturing Scale Production Manufacture at the scale typically encountered in a facility intended for product production for marketing.

MANUFACTURER'S WORKING CELL BANK (MWCB) OR WORKING CELL BANK (WCB) A cell bank derived by propagation of cells from MCB under defined conditions and used to initiate production cell cultures on a lot-by-lot basis.

Mass spectrometry An analytical technique in which ions are separated according to their mass-to-charge ratios.

MASTER CELL BANK (MCB) A bank of a cells from which all subsequent cell banks used for production will be derived. The MCB represents a characterized collection of cells derived from a single tissue or cell.

Melting temperature (*T*m) The temperature at which a double-stranded DNA or DNA–RNA molecule denatures.

Messenger RNA (mRNA) The ribonucleic acid molecule transcribed from DNA that carries the codons specifying the sequence of amino acids in a protein.

Microinjection Introduction of DNA into the nucleus or cytoplasm of a cell by insertion of a microcapillary and direct injection.

Micro RNAs (miRNAs) Short RNA molecules synthesised as part of the RNA interference mechanism.

miRNA See *Micro RNAs*.

Molecular cloning Alternative term for gene cloning.

Monomer The unit that makes up a polymer. Nucleotides and amino acids are the monomers for nucleic acids and proteins, respectively.

Multiple cloning site (MCS) A short region of DNA in a vector that has recognition sites for several restriction enzymes.

Mutant An organism (or gene) carrying a genetic mutation.

Mutation An alteration to the sequence of bases in DNA. May be caused by insertion, deletion, or modification of bases.

Northern blotting Transfer of RNA molecules onto membranes for the detection of specific sequences by hybridisation.

N terminus Amino terminus, defined by the -NH_2 group of an amino acid or protein.

Nuclease An enzyme that hydrolyses phosphodiester bonds.

Nucleoside A nitrogenous base bound to a sugar.

Nucleotide A nucleoside bound to a phosphate group.

Nucleoid Region of a bacterial cell in which the genetic material is located.

Nucleus Membrane-bound region in a eukaryotic cell that contains the genetic material.

Oligo Prefix meaning few, as in oligonucleotide or oligopeptide.

Oligo(dT)-cellulose Short sequence of deoxythymidine residues linked to a cellulose matrix, used in the purification of eukaryotic mRNA.

Oligomer General termfor a short sequence of monomers.

Oligonucleotide A short sequence of nucleotides.

Open reading frame (ORF) A series of codons that is or could be a gene.

Operator Region of an operon, close to the promoter, to which a repressor protein binds.

Operon A cluster of bacterial genes under the control of a single regulatory region.

Origin of replication The specific position on a DNA molecule where DNA replication begins.

Palindrome A DNA sequence that reads the same on both strands when read in the same (*e.g.* 5'→3'_) direction. Examples include many restriction enzyme recognition sites.

PARENT CELL BANK A few vials consisting of cells from which the Master Cell Bank was derived. Parental Cells may be manipulated to derive a cell substrate with desired characteristics.

PASSAGE LEVEL The number of times, since establishment from a primary cell culture, a culture has been split or re-seeded.

PCR See *Polymerase chain reaction*.

Phosphodiester bond A bond formed between the 5'phosphate and the 3' hydroxyl groups of two nucleotides.

Pilot-Plant Scale The production of the drug substance or drug product by a procedure fully representative of and simulating that to be applied at manufacturing scale. The methods of cell expansion, harvest, and product purification should be identical except for the scale of production.

Plasmid A circular extrachromosomal element found naturally in bacteria and
some other organisms. Engineered plasmids are used extensively as vectors
for cloning.

Polyacrylamide A cross-linked matrix for gel electrophoresis (q.v.) of small fragments of nucleic acids, primarily used for electrophoresis of DNA. Also used for electrophoresis of proteins.

Polyadenylic acid A string of adenine residues. Poly(A) tails are found at the 3'
ends of most eukaryotic mRNA molecules.

Polycistronic Refers to an RNA molecule encoding more that one protein. Many bacterial operons are expressed *via* polycistronic mRNAs.

Polylinker See *Multiple cloning site*.

Polymer A long sequence of monomers.

Polymerase chain reaction (PCR) A method for the selective amplification of DNA sequences. Several variants exist for different applications.

Polynucleotide A polymer made up of nucleotide monomers.

Polypeptide A chain of amino acid residues. *Cf. Protein.*

POPULATION DOUBLING LEVEL The number of times, since establishment from a defined point in the history of a cell substrate (often the primary cell culture), a culture has doubled in number of cells.

Post-translational modification Modification of a protein after it has been synthesised. An example would be the addition of sugar residues to forma glycoprotein.

Potency The measure of the biological activity using a suitably quantitative biological assay (also called potency assay or bioassay), based on the attribute of the product which is linked to the relevant biological properties.

PRIMARY CELLS: Cells placed into culture immediately after an embryo, tissue, or organ is removed from an animal or human and homogenized, minced, or otherwise separated into a suspension of cells.

Primary transcript The initial, and often very large, product of transcription of a eukaryotic gene. Subjected to processing to produce the mature mRNA molecule.

Primer A short single-stranded oligonucleotide which, when attached by base pairing to a single-stranded template molecule, acts as the start point for complementary strand synthesis directed by a DNA polymerase enzyme.

Primer extension Synthesis of a copy of a nucleic acid from a primer. Used in labelling DNA and in determining the start site of transcription.

Probe A labelled molecule used in hybridisation procedures.

Prokaryotic The property of lacking a membrane-bound nucleus (*e.g.* in bacteria such as *E. coli*).

Promoter DNA sequence(s) lying upstream from a gene, to which RNA polymerase binds.

Protease Enzyme that hydrolyses polypeptides.

Protein A condensation (dehydration) heteropolymer composed of amino acid residues linked together by peptide bonds to give a polypeptide.

Radiolabelling Short for radioactive labelling; method used to incorporate radioactive isotopes into biological molecules. An example is labelling nucleic acids with 32P-dNTPs to prepare high-specific-activity probes for use in hybridisation experiments.

Reading frame The pattern of triplet codon sequences in a gene. There are three reading frames, depending on which nucleotide is the start point. Insertion and deletion mutations can disrupt the reading frame and have serious consequences, as often the entire coding sequence becomes nonsense after the point of mutation.

Recombinant DNA molecule A DNA molecule created in the test tube by ligating together pieces of DNA that are not normally contiguous.

Recombinant DNA technology All of the techniques involved in the construction, study
and use of recombinant DNA molecules.

Recombinant protein A polypeptide that is synthesized in a recombinant cell as the result of expression of a cloned gene.

Relaxed Refers to a plasmid with a high copy number of perhaps 50 or more per cell. (2) The non-supercoiled conformation of open-circular DNA.

Replication Copying the genetic material during the cell cycle. Also refers to
the synthesis of new phage DNA during phage multiplication.

Replicon A piece of DNA carrying an origin of replication.

Resin A chromatography matrix.

Restriction enzyme An endonuclease that cuts DNA at sites defined by its recognition sequence.

Restriction fragment A piece of DNA produced by digestion with a restriction enzyme.

Retrovirus A virus that has an RNA genome that is copied into DNA during the infection.

Reverse transcriptase An RNA-dependent DNA polymerase found in retroviruses, used *in vitro* for the synthesis of cDNA.

Ribonuclease (RNase) An enzyme that hydrolyses RNA.

Ribonucleic acid (RNA) A condensation heteropolymer composed of ribonucleotides.

Ribosomal RNA (rRNA) RNA that is part of the structure of ribosomes.

Ribosome-binding site A region on an mRNA molecule that is involved in the binding of ribosomes during translation.

RNA processing The formation of functional RNA from a primary transcript. In mRNA production this involves removal of introns, addition of a 5'cap, and polyadenylation.

RT-PCR Reverse transcriptase (transcription) PCR, where a cDNA copy of mRNA is made and then amplified using PCR.

Saccharomyces cerevisiae Unicellular yeast (baker's yeast, also known as budding yeast) that is extensively used as a model microbial eukaryote in molecular studies. Also used in the biotechnology industry for a range of applications, as well as in brewing and bread-making.

Screening Identification of a clone in a genomic or cDNA library by using a method that discriminates between different clones.

Selection Exploitation of the genetics of a recombinant organism to enable desirable, recombinant genomes to be selected over non-recombinants during growth.

Simian virus 40 (SV40) A mammalian virus that has been used as the basis for a cloning vector.

Somatotropin Growth hormone

Southern blotting Method for transferring DNA fragments onto a membrane for detection of specific sequences by hybridisation.

Sticky ends See *Cohesive ends*.

Strong promoter An efficient promoter that can direct synthesis of RNA transcripts at a relatively fast rate.

Structural gene A gene that encodes a protein product.

Taq **polymerase** Thermostable DNA polymerase from the themophilic bacterium*Thermus*
aquaticus. Used in the polymerase chain reaction.

Template A single-stranded polynucleotide (or region of a polynucleotide) that directs
synthesis of a complementary polynucleotide.

Terminator The short nucleotide sequence, downstream of a gene, that acts as a signal
for termination of transcription.

Terminal transferase An enzyme that adds nucleotide residues to the 3_ terminus of an oligo- or polynucleotide.

Tetracycline (Tc) A commonly used antibiotic.

Thermal cycler Heating/cooling system for PCR applications. Enables denaturation, primer binding, and extension cycles to be programmed and automated.

Thermus aquaticus Thermophilic bacterium from which *Taq* polymerase is purified. Other bacteria from this genus include *Thermus flavus* and *Thermus thermophilus*.

Thymine (T) Nitrogenous base found in DNA only.

Thymidine Nucleoside composed of deoxyribose and thymine.

Transcription (TC) The synthesis of RNA from a DNA template.

Transfer RNA (tRNA) A small RNA that carries the anticodon and the amino acid residue required for protein synthesis.

Transformant A cell that has been transformed by exogenous DNA.

Transformation The process of introducing DNA (usually plasmid DNA) into cells. Also used to describe the change in growth characteristics when a cell becomes cancerous.

Transgenic An organism that carries DNA sequences that it would not normally have in its genome.

Translation (TL) The synthesis of protein from an mRNA template.

Uracil (U) Nitrogenous base found in RNA only.

Vector A DNA molecule that is capable of replication in a host organism and can act as a carrier molecule for the construction of recombinant DNA.

Virus An infectious agent that cannot replicate without a host cell.

Western blotting Transfer of electrophoretically separated proteins onto a

membrane for probing with antibody.

WCB (Working Cell Bank) The Working Cell Bank is prepared from aliquots of a

homogeneous suspension of cells obtained from culturing the MCB under defined culture condition.

www.ingramcontent.com/pod-product-compliance
Lightning Source LLC
Chambersburg PA
CBHW070833180526
45168CB00002B/827

* 9 7 8 1 5 0 7 7 1 2 0 9 2 *